CAMBRIDGE LIBRARY C(

Books of enduring scholarly value

Life Sciences

Until the nineteenth century, the various subjects now known as the life sciences were regarded either as arcane studies which had little impact on ordinary daily life, or as a genteel hobby for the leisured classes. The increasing academic rigour and systematisation brought to the study of botany, zoology and other disciplines, and their adoption in university curricula, are reflected in the books reissued in this series.

Memorials of Sir C.J.F. Bunbury

Sir Charles James Fox Bunbury (1809–86), the distinguished botanist and geologist, corresponded regularly with Lyell, Horner, Darwin and Hooker among others, and helped them in identifying botanical fossils. He was active in the scientific societies of his time, becoming a Fellow of the Royal Society in 1851. This nine-volume edition of his letters and diaries was published privately by his wife Frances Horner and her sister Katherine Lyell between 1890 and 1893. His copious journal and letters give an unparalleled view of the scientific and cultural society of Victorian England, and of the impact of Darwin's theories on his contemporaries. Volume 6 covers the years 1869–72. Bunbury was less active in scientific matters than before, but still reading copiously, carrying on a busy social life, and developing the gardens of his home in Suffolk. There are many comments on current affairs and politics.

Memorials of Sir C.J.F. Bunbury

VOLUME 6: LATER LIFE PART 2

EDITED BY
FRANCES HORNER BUNBURY
AND KATHARINE HORNER LYELL

CAMBRIDGE UNIVERSITY PRESS

Cambridge, New York, Melbourne, Madrid, Cape Town,
Singapore, São Paolo, Delhi, Tokyo, Mexico City

Published in the United States of America by Cambridge University Press, New York

www.cambridge.org
Information on this title: www.cambridge.org/9781108041171

© in this compilation Cambridge University Press 2011

This edition first published 1891
This digitally printed version 2011

ISBN 978-1-108-04117-1 Paperback

MEMORIALS

OF

Sir C. J. F. Bunbury, Bart.

EDITED BY HIS WIFE.

THE SCIENTIFIC PARTS OF THE WORK REVISED BY
HER SISTER, MRS. LYELL.

LATER LIFE.

Vol. II.

MILDENHALL:
PRINTED BY S. R. SIMPSON, MILL STREET.
MDCCCXCI.

1869.

JOURNAL.

A beautiful, bright, clear day, with sharp frost. 1869.
I spent the morning quietly with the Arthur
Herveys at Ickworth. Fanny arrived soon after
twelve, much better. We stayed till after luncheon
with the dear Herveys, then we returned home.

The labourers' supper party.

Again I feel myself impelled to begin this new
year with an expression of my deep, humble, and
heartfelt gratitude to Almighty God for all his
goodness to me, and for the innumerable blessings
which I enjoy. I look back on the year 1868 as well
as the one which preceded it, as periods of unmixed,
unclouded happiness; clouded, at least, only by
the shadows which the sorrows of others cast over
us; and of these there has not been much to touch
us nearly, except poor Scott's great misfortune in
the loss of his excellent wife, near the beginning of
last year. What this new year may bring us—who
can tell? but whatever it may be, I can never
cease to be grateful for the blessings which have
been showered upon me.

Read to Fanny the " Beggars' Opera," which I

1869. have not looked at for many years; and I cannot help wondering how any respectable audience could ever endure to see it acted.

———

January 3rd, Sunday.

We went to Church and received the Communion. Walked with Fanny to see the new north lodge.

———

January 7th, Thursday.

We all dined with Lady Cullum. Met Mr. and Mrs. Joseph Hooker, Milner Gibson, Mrs. Abraham, James Bevans, and others—a pleasant party.

Joseph Hooker talked to me of a prospect he has in view, which I think would be useful to botanists—that there should be a herbarium and botanical library *kept together* for constant reference, at the British Museum; the herbarium to be a specially typical one, of select characteristic specimens, carefully named, so as to be as convenient as possible for consultation and for the determination of species. Such a herbarium, he says, might be formed from the duplicates of Kew, and from the more modern part of the collections in the British Museum. The old herbaria in the Museum, such as Sloane's, might be removed to Kew.

The great inconvenience, at present, of the botanical collections in the British Museum is, that the books connected with the subject are all in a different department, and under another charge. Hooker says that the Abyssinian expedition has not added much to our knowledge of the natural history of that country; there was little time

or convenience for observing or collecting; and 1869.
much was already known, through the researches of
some French and German Naturalists, who had
resided for years in that country. Its botany, in
particular, had been well studied by Schimper, who
is still living there, and declined to come away with
the army. Hooker says, it is by no means a rich
botanical country.

January 8th, Friday.

Paid the bill for the new organ. William Napier
has published in *The Times*, a very spirited account
of the fighting at Malaga, from a letter of his wife,
who was in the thick of it.

LETTER.

Barton,
January 9th, '69,

My Dear Edward,

It is rather late in the day to wish you a
happy new year, but nevertheless, I do wish it you
with all my heart and many more of the like. We
have been very quiet since we returned home. It
is, indeed, an extraordinary winter, the farmers
rejoice in it, for the remarkably *open* weather is so
favourable to the grass and the green crops, that it
makes up in some degree for the drought of the
summer and autumn; it is good for the poor
also.

We have Snowdrops and yellow Aconites quite in

1869. flower in the grounds, the wood-pigeons are cooing
and little birds singing.

Lady Cullum is flourishing. We all (Clement
and Herbert included) dined with her the other day,
and met Joseph and Mrs. Hooker, as well as Milner
Gibson and his son. Fanny was not well enough
on New Year's Eve to go with me to the Arthur
Herveys party, where they had a very pleasant
dinner, and afterwards charades capitally performed
by the young people of the two houses. Sarah
and her brothers and sisters, Lady Mary, Lord
John and Lord Francis. The words were—*man-
slaughter* and *Penelope*. Lady Mary looked really
beautiful as Penelope : and as her brother was
Ulysses, they were not reduced to *shaking hands* on
recognition, as you and Susan Horner were when
you acted the same long ago at Mildenhall. While
we were alone here I read "The Spanish Gypsy"
aloud to Fanny, and we were delighted with it : I
especially so ; it is the most beautiful new poem I
have read for a long time. Now I am reading to
her some of Lord Campbell's "Lives of the
Chancellors." She is also reading Lord Hervey's
"Memoirs," with which she is much entertained :
and I am reading Raleigh's "History of the
World," which, however, as far as I have yet gone
disappoints me : it is excessively lengthy, and does
not appear to me at all philosophical.

We expect the MacMurdos the day after to-
morrow, to stay with us, I believe, till near the end
of the month : and early in February I hope the
Kingsleys come to us. I am feeling very thankful

for my good fortune in that no Judges are coming 1869. into Suffolk to try election petitions.

Fanny sends you her love, and says she will write to you soon,

Ever your very affectionate brother,

CHARLES J. F. BUNBURY.

JOURNAL.

January 11th. Monday.

Last Saturday morning, the 9th, about 11.30 a.m. there was felt in many parts of this parish and neighbourhood a shock which was taken by various people for that of an earthquake. I cannot say that I or Fanny felt it, but it was strongly felt in Scott's house, at the Home Farm, and we have heard of it from a variety of places in this and the neighbouring parishes: in some places the shock was so strong that people were quite frightened and thought their houses were about to fall. It was felt by the people working in the open fields, and it seemed, they said, to come from under their feet. It was, doubtless, the effect of some explosion of powder mills or of some manufactory of that sort, of which the newspapers have not yet brought us the account.

I have read to Fanny in the evenings, the "Life of Lord Harcourt," in Lord Campbell's "Lives of the Chancellors:"—very agreeably written.

1869.

The MacMurdos, Montagu, Susan, Kate and Mimie arrived.

———

It is very strange that nothing has come out to explain the *shocks* experienced on the 9th. No explosion is recorded in the Newspapers. The shocks are mentioned, and attributed to an earthquake, in the *Times* of the 12th, but *only* in letters from Suffolk.—I can find no notice of their having been felt elsewhere. Captain Horton, at Livermere, felt the shock distinctly, and thought decidedly that it was an earthquake : and he has experienced earthquakes in countries where they are more usual. Yet it seems very unlikely that a real earthquake should have been confined to so small a district.

MacMurdo tells me that our iron war steamers have come triumphantly out of the trials to which they have been exposed, in the storms of this winter. They are found to be more buoyant, and to pass more lightly over the heaviest seas than wooden ships, and to suffer much less from the straining occasioned by such heavy usage.

———

We walked with Scott round the grounds. Inspected operations and gave directions. Heard of the death of (Lord Arthur Hervey's sister) Lady Georgina Grey. Very sorry.

———

January 20th. Wednesday. 1869.

General and Mrs. Eyre came to luncheon and stayed late.

————

January 21st. Thursday.

Lady Parker and her sister Miss Barnardiston, Mrs. and Miss Loch and the two young Angersteins arrived for the Bury Ball, to which Fanny and all the party, except George Napier and myself went.

————

January 25th. Monday.

We walked round the grounds and labelled several trees. I examined the fruit of Cobæa. Sir George Young went away. Edward arrived—also Kate Hervey.

LETTER.

Barton,
January 25th, 69.

My Dear Katharine,

I thank you much for your letter. We had a pleasant little party for the ball : besides the MacMurdos, there were Mrs. and Miss Loch, Lady Parker and her sister Miss Barnardiston, Sir George Young and two young Angersteins. The ball was said to be a good one, although all the Ickworth party were kept away by the sudden death of poor Lady Georgina Grey.

I am glad to hear that you have made so much progress with your book on Ferns, which does

1869. infinite honour to your zeal and perseverance.   I
shall be very glad to look through the African
portion whenever you like to send it to me.

You may perhaps have heard that Charles
Kingsley is going out with his daughter on a visit to
Trinidad this next spring, and this has set me upon
picking up all the information within my reach upon
West Indian botany.   It will be a most enviable
excursion.   I cannot imagine any one better
qualified to enjoy and profit by it than Kingsley:
and Rose will be a companion worthy of him.

Fanny is reading Burnet's "History of his Own
Time," and to keep up a connexion between our
studies, I am reading the same period in Hallam's
"Constitutional History."   This I read, of course,
many years ago, and I admire it now more than
ever.

I have not time to write more.

Ever your very affectionate brother,
CHARLES J. F. BUNBURY.

JOURNAL.

January 26th.   Tuesday.

George Napier went away.   Sarah Hervey with
her two young sisters came and spent the afternoon
with us, and was very agreeable.

I am glad to find from the papers, that Henry
Bruce has come into Parliament without opposition
for Renfrewshire.

I took a walk with Montagu MacMurdo and
Fanny.   We dined with the Abrahams at Risby:—
Bishop Abraham* a very interesting man.

————

January 28th. Thursday.

The MacMurdos went away.   Montagu Mac-
Murdo is a man for whose abilities and character I
have a high admiration, and in whose society I take
great pleasure.   He is cheerful, animated, genial
and very agreeable in conversation.   He has seen
and observed a great deal, both of men and things
in many various countries, and has brought a clear
and active mind to work on what he has observed.
His stories of what he has seen and done are
admirably clear and vivid.   I never heard anything better told than his incidents which he related
to us while here:—the one, his success over a
horde of Beloochees who attacked him while with
only twelve troopers under his command, he was
driving off cattle through a narrow pass in the Hill
country; the other, Sir Charles' danger while
passing on an elephant over a weak bridge of boats
across the Indus.

We went to Professor Churchill Babington's
wedding breakfast at the Angel.—A splendid affair
and very well managed.   I sat by Lady Cullum.
Some of the speeches good, especially Mr. Babington's and Professor Lightfoot's.†

* Suffragan Bishop of Lichfield; an intimate friend of Bishop
Selwyn.—(F. J. B.)

† Afterwards Bishop of Durham.

1869.

Clement went away. We labelled some more trees.

* * *

We went to the Vicarage to see the progress of the alterations.

* * *

Mr. and Mrs. Chapman and Mr. Stopford Brooke came to luncheon and to see our pictures. We went to Ickworth to see the Arthur Herveys, who are going to London to-morrow; had a pleasant little chat with them.

* * *

My 60th birthday. Thanks to God.

* * *

LETTER.

Barton,
February 4th, 1869.

My Dearest Cissy,

I thank you heartily for your truly kind and loving letter, which has gratified me exceedingly by its warm expressions of cordial affection in the sincerity of which I feel the most undoubting confidence. I do indeed, my dearest sister, feel most entirely assured of your and dear Henry's constant love and kindness to me, of which I have had so many proofs; and I think I need hardly tell you that I most cordially return your affection, which I

reckon among the greatest blessings of my life. I 1869.
have indeed reached the respectable, not to say
venerable age of sixty ; and I feel very thankful to
the Almighty that I have been permitted to arrive
at this age in good health of body and mind, and in
the enjoyment of such an unusual measure of prosperity and of so many blessings. It is a great comfort that Fanny has latterly been very tolerably well.

This evening our servants have a dance. In the
morning the Sutton* children very kindly brought
me a little bouquet, and I have had very gratifying
expressions of regard and good wishes from Mr.
Phillips and from Scott.

I was delighted by your account of dear little
Willy, and especially of his Whiggism, and I trust
he will go on perfectly to your satisfaction at school.
Very happy to hear that George is well. Pray thank
Henry and dear Emmy very much in my name for
their kind letters which I have received this evening,
and with very much love from both of us to you and
all yours, believe me ever,

Your very affectionate brother,
CHARLES J. F. BUNBURY.

P.S.—Very many and cordial thanks for my father's
beautiful little sketch, which I value exceedingly.

———

Barton,
February 5th, 1869.

My Dear Katharine,
I thank you very much for your kind letter and
good wishes on my birthday—on my attaining the

* The Suttons lived at the Cottage.

1869. *venerable* age of sixty. I have very great reason to be thankful for my good health, and the numberless blessings that I enjoy—above all for having my wife by my side, and in good health. I do not feel at all inclined to mourn my departed youth, but feel as happy as ever I did, and have nothing to wish but that the years which may remain to me may be well spent.

We were out all yesterday morning enjoying the delightful weather (it was perfect spring) and the beauty of the flowers. Crocuses of three species, snowdrops of two, Violets, Hepaticas, and a lovely little blue Scilla, are profusely in blossom, and primroses are beginning ; also the delicate little Cynoglossum omphalodes. Pyrus japonica coming into flower. The grass is as green as in April, and the birds singing delightfully. It is a wonderful season.

Ever your affectionate brother,
CHARLES J. F. BUNBURY.

JOURNAL.

February 6th, Saturday.

Our dear friends, the Kingsleys with their two charming daughters, Rose and Mary, arrived. We had a very pleasant evening and conversation with them.

February 7th, Tuesday.

We all went to morning Church, where Kingsley

read the service. and Mr. Smith preached an 1869. excellent sermon. I had a very pleasant walk round the grounds with Kingsley.

Kingsley went early to Cambridge to his lectures and returned in the evening.

I had a very pleasant walk with the girls.

Mr. and Mrs Chapman and two Miss Erskines* and Mr. Percy Smith dined with us.

Spent the evening very pleasantly looking over old books with the Kingsleys.

Spent much of the day very pleasantly looking over dried plants with the Kingsleys.

Mr. and Lady Susan Millbank, Mr. and Mrs. Abraham and Lady Cullum dined with us.

Mr. and Lady Mary Powys arrived. A fall of snow at night.

A white frost in the morning. The afternoon very fine. Looked over dried plants with Kingsley. All the party except me went over to see Hardwick. Lord John Hervey and Sir George Young arrived. A very pleasant talk with Kingsley in the evening.

* The daughters of Judge Erskine and grand-daughters of Lord Erskine.

February 14th, Sunday.

We all went to morning Church — Kingsley preached an admirable sermon. He read us another excellent one in the evening.

We looked over prints and over the curiosities in the little cabinet.

———

February 15th, Monday.

Mr. and Lady Mary Powys and Sir George Young went away. Rev. John Hervey arrived late. I had a very pleasant walk with the two dear Kingsley girls and Lord John Hervey. We looked over e ngravings in the evening.

———

February 16th, Tuesday.

Had a very pleasant walk with Fanny and Mr. Kingsley about the park.

We had a large dinner party—Lady Cullum, Wilsons, Rodwells, etc.

———

February 18th, Thursday.

We planted a tree for dear Mary Kingsley. I looked over some dried plants with Kingsley. Lady Susan Milbank arrived.

On my mentioning that peculiar expression in the 21st chapter of Acts, " we took up our carriages," Kingsley said that though he has been in the habit of reading much of the English writers of that age, he has never met with another instance of the word "carriages" used in that sense, namely as equivalent to *baggage*. He seemed to think that it 1869. was a mere mistake. Our version of the New Testament, he said, is extremely unequal, some parts very good, some showing great ignorance.

Kingsley told me that Prince Albert many years ago, tried to naturalize both the Capercalli and the Red Grouse in the heath country about Bagshot. The Grouse soon dispersed and disappeared, but a few of the Capercalli still remain, and are seen from time to time; he himself (C. K.) saw one there not long ago. But they have not increased, and the reason of this, as well as of the disappearance of the Red Grouse, is, he believes, the want of sufficient food. It is quite idle to think of naturalizing any kind of bird where they cannot have an ample supply of food suited to them. Heath alone is not sufficient for the Grouse. On the famous Grouse-moors of Yorkshire, he found that the growth consisted in almost as large a proportion of Bilberries, Crowberry and other berry-bearing plants as of Heath. The supposed *epidemic* among the Grouse a few years ago, was really, he believes, occasioned by insufficiency of food, their numbers having increased out of due proportion to the means of subsistence.

Kingsley remarked that the professorial system, which in the middle ages prevailed in all the Universities of Europe, and which still prevails in those of the Continent, has in England been superseded by the monastic system of the Colleges. Our Colleges are very rich, our Universities very poor. Kingsley said, in almost all cases of unhappy

1869. marriages which had come under his observation,
the husband had been an *idle* man, without a
profession and without any earnest special pursuit to
supply the place of a profession.

_____

February 19th. Friday.

A most beautiful day—perfect spring. We were
out much of the morning with Scott and Allan
planning alterations.

_____

February 20th. Saturday.

Charles and Mary Lyell, the Hookers and Capt.
Brine arrived. Mr. and Mrs. P. Smith and Mr.
Abraham dined with us.

_____

February 21st. Sunday.

We all went to Church in the morning. Mr.
Smith's sermon excellent. I had a pleasant walk
back with Mrs. Kingsley, and another in the afternoon with Charles and Mary.

Kingsley said (and I quite agree with him) that
one of the most dangerous classes in this country is
the lower sort of literary men, especially many
of those who are employed on newspapers and
periodicals. The most prevalent of all vices in the
present day, is envy: and much of what is called
the democratic spirit is nothing else than mere
envy.—(C. K.)

Kingsley is inclined to be favourable to a con-
scription (on something like the Prussian system) in
England; he thinks that most young men would

be rather the better for three years of military 1869. discipline and training, between the period of school and that of the regular business of life. Of all the elements of society in England, the most stable is the class of country gentlemen.—(C. Kingsley.)

We were talking of the Balaclava charge; Kingsley said the moral effect of it was worth all the sacrifice of life. It might be an error in a purely and technically military view : but only in that view.

Kingsley has been told (I believe by his friend Mr. Gordon,* the governor of Trinidad), that the practice of *Obi* (pagan magical rites) is still frequent among the professedly Christian Negroes of the West Indies, and that (as among the *veneficæ* of ancient Rome) this magic is generally combined with the practice of poisoning. In particular, that the Obi men have a practice of sharpening their thumb nail and anointing it with a deadly poison, so that any scratch with it is fatal. Humboldt mentions the same practice among the Otomaques of the Orinoco.

————

February 22nd. Monday.

Kingsley left us in the morning.—Dr. and Mrs. Hooker after luncheon.

Charles Lyell tells me that remarkably perfect fossil specimens of Taxodium distichum have been lately brought from Spitzbergen, supposed to be of tertiary age : leaves, twigs, fruits and seeds, so well

---

* Afterwards Sir Arthur Gordon.

1869.   preserved that it is thought there can be no doubt
about the generic and specific identity of the fossil
with the recent deciduous Cypress.

February 23rd.   Tuesday.

Very cold.   We were all photographed in the
conservatory.   Dear Mrs. Kingsley and her
daughters left us.   I must fully confirm here what I
have more than once written before of my love and
admiration for the Kingsleys.   The two girls, Rose
and Mary, are thoroughly worthy of such a father
and Mother.

Kingsley's thoughts are very intent upon a
delightful expedition which he has in prospect : he
and Rose are going out to Trinidad with the
governor Mr. Gordon, who is an intimate friend of
his, to spend six weeks or two months there.   With
his intimate knowledge of all the history of the
West Indies, his passionate love of nature, his
knowledge of various branches of natural history,
his ardour and enthusiasm, and with such a companion as Rose, it will be as enjoyable an expedition
as can be imagined.   I trust both of them will
preserve their health.

Kingsley and I spent many hours together in
looking over plants in my herbarium as he was
anxious to make himself familiar with the prevailing
types of tropical American vegetation.   He has
been assiduously studying Lindley's Vegetable
Kingdom, for which he has a very just admiration.

February 24th.   1869.

Scott gave a good report of the Mildenhall rent audit.

———

February 25th, Thursday.

Walked with Charles Lyell to Barton Mere ; Mr. H. Jones was absent in London, but his steward showed us the principal objects which have been found at the bottom of the mere, in the operation of cleaning out and deepening it. There is in particular a large collection of fragments of bones and horns, which have been examined and marked by Boyd Dawkins; the most numerous of them are of red deer and horse, but there are also several of the Urus. (? Bos primigenius). There were also found a quantity of posts or pales of wood, shaped by art.

———

February 27th.

Charles and Mary Lyell left us. Both, I am happy to say, appear to be in pretty good health, but Charles shows some signs of old age.

———

February 28th.

This month (February) ends badly as to weather, a most bitter and violent wind prevailed, and much snow fell during the day, but did not lie long. During nearly the whole of the month, the weather has been most remarkably mild ; quite spring-like ; many days really beautiful. The earliness of the season has been quite extraordinary, and I do not remember ever to have seen the garden so gay with flowers in February. At the beginning of the

1869. month, snowdrops, crocuses of two kinds (the common yellow and the cloth of gold) violets, hepaticas, and a beautiful little blue squill, were in full blossom. The wild crocuses in the park, were in flower by the 5th; pyrus japonica a few days later; the almond tree on the 16th; the peach (against a wall) on the 17th, the daffodil on the 19th; ribes sanguineum on the 26th. The weeping willow quite in leaf by the 14th, and the lilac beginning to expand its leaves by the 20th. I never saw the pyrus japonica, or the almond, in such beauty as this year.

LETTER.

Barton,
March 1st, 1869.

My Dear Joanna,

I owe you and dear Susan very many thanks for your kind letters and good wishes on the occasion of my birthday, and I feel very sure that, far as we now are separated in local habitation, I am in no danger of being forgotten by such kind sisters as you have always shown yourselves. I am extremely glad that you are so comfortably established in Florence, and find yourselves so happy there, and have so many pleasant acquaintances; and it must have been very pleasant to find yourselves so cordially welcomed on your return by your former friends, and to feel yourselves so much at home in beautiful Florence.

We have had a great pleasure in a visit of a fortnight from the Kingsleys, Mr. and Mrs.

Kingsley and their two daughters, whose society is 1869.
always a special delight to me; I think I like and
love them more and more every year; Mrs. Kingsley
is quite as much superior to the generality of people
one meets as her husband. Then we had the pleasure
also of dear Charles and Mary Lyell, staying with
us a week, and some other pleasant people from
time to time. And now we are again alone together
and hope to get through a good deal of reading and
of quiet work.

I hope you read Lanfrey; I have just finished
the third volume; it is one of the most remarkable
specimens of idol-breaking that I have ever met
with. The way in which he strips the gilding off
the great Emperor and exhibits him in his true
detestable character, is most striking, and severe as
his exposure is, it is entirely founded upon
Napoleon's own correspondence. I have been con-
stantly reminded in reading this book, of Byron's
fine ode, beginning—

" 'Tis done—but yesterday a King."

I am afraid we can hardly hope that the "spell
upon the minds of men" is effectually broken; but
it is a good thing that the wickedness of the greatest
of conquerors should be thoroughly exposed. And
I must add that Thiers, as a historian, is as much
damaged by Lanfrey as his hero is. I have also
lately read Colonel Chesney's Lectures on the
Waterloo campaign; he seems to be the first
English writer who has done full justice to the
Prussians, and clearly shown how much of the

1869. final success depended on Gneiseneau's decision
after the battle of Ligny.   Fanny has for some time
been reading Burnet's "History of His Own Time,"
and I am just beginning it, having lately read
through the reigns of Charles and James the Second
in Hallam's "Constitutional History."

Much love to Susan, to whom I will write as soon
as I have anything to say.

Ever your very affectionate brother,
CHARLES J. F. BUNBURY.

JOURNAL.

March 2nd, Tuesday,

My Barton rent audit.   After it I had luncheon
with my tenants, Mr. Percy Smith being of the
party.

Scott tells me that Mr. Darkin, the builder,
cultivates vines on his land in Bury, for the purpose
of making *wine*, and actually does make very drink-
able wine from them.   He cuts them down into
low bushes, as it is done in France and Germany.

March 3rd, Wednesday

The weather has become more wintry ; therm.
last night down to 19 deg.   Yesterday, Fanny and
I stood for some time watching a creeper (Certhia
familiaris) running or rather gliding, up the trunk of
an old ash-tree in the park.   The movement is very
curious and pretty ; the bird seems to glide like a
mouse, in a wavy or zigzag line up the tree, stopping

at short intervals.   Its climb ended in a hollow 1869. broken branch, high up the tree, where it probably may be making his nest.   It may very likely be a common bird here, but one seldom gets a good sight of it.

LETTER.

Barton,
March 4th, 1869.

My Dear Katharine,

I have gone carefully through your Asiatic Fern lists.   I admire your industry and rejoice that you have made such progress in your work.   The revision of your lists has led me to look a good deal into the " Synopsis Filicum ;"   it is a useful book, but I cannot at all make out on what principles Mr. Baker has proceeded in settling the species, or why he has gone so far in reducing species, and not gone much further.   As to the genera, with a few excep-tions, they are quite unsatisfactory to me—utterly artificial, and I cannot help thinking that botanists have not yet discovered the true principles on which the genera of Ferns are to be founded.   What Joseph Hooker says on this subject in the "Flora of New Zealand" appears to me more sound than anything else I have read upon it.   No doubt, as there is such confusion between the many systems proposed, it may be convenient to adopt the genera of the "Synopsis" provisionally, as a make-shift ; but I think they are hardly at all in advance of Swartzy's method in 1806.

1869.   I suppose this sharp cold which has come at last may be good for people in general, but it is very disagreeable to me.

With much love to Harry, Rosamond, Leonard and the rest, believe me,

Ever your very affectionate brother,

CHARLES J. F. BUNBURY.

LETTER.

Barton,
March 10th, 1869.

My Dear Henry,

Very many thanks for your letter. All you say about your boys is very interesting to me, as everything must be which so deeply concerns you and dear Cissy, besides that, I love the dear little fellows for their own sakes. I trust you need have no serious anxiety about Henry, I mean as to essentials, but it is certainly a very serious question to be considered what will be the best profession for him, in case he does not pass for the navy.

I think the choice of a profession for a gentleman, in these times, is one of the most difficult of problems.

We must bear in mind that, in all human probability, if he lives, he will one day be the owner of these estates: therefore, it appears to me, his education should be emphatically that of a *gentleman* —not of an *idle* man by any manner of means, but

of one who should be a cultivated as well as a 1869.
practically useful man, qualified to play his part
well in the best society, as well as to understand
and manage business. I do not for a moment wish
to lay down the law, I only mention what occurs
to me. With respect to the army, whether the line
or the guards, I confess I cannot help thinking it.
rather a dangerous profession, in time of peace, for a
young man who has either wealth or the prospect of
it : the temptations to idleness are so great. But
this objection does not apply, I should suppose, to
the engineers, and with Henry's peculiar turn for
everything of machinery and mechanical and
chemical science, I should think *that* would be the
profession best suited to him. The examinations,
no doubt, are hard, but they might suit his turn of
mind better than those required elsewhere. The
difficulty always great of choosing a profession,
seems to be increased just now by the violent fit of
reduction and retrenchment which has seized upon
every department of the administration : and this,
as you say, makes it more likely that he will not be
allowed to pass at his next examination for the
navy. I hardly suppose that one of the more
especially *studious* professions would suit him.

I have written at great length, but I wished you
to see that I feel all the importance and interest of
the question. I hope that in whatever decision you
may come to, as to the best course for your boy's
welfare, you will not be checked by consideration of
expense : I shall be most happy to help you, and
indeed I should be very sorry not to share in the

1869. expense, whether of giving him the best education, or of launching him in a profession.

One thing more: supposing Henry fails to pass for the navy, have you quite decided that the best plan will be to leave him at his present preparatory school? or might it not be a good thing to send him to one of the great schools?—I should like to know what you think of this. I suppose, indeed, that if Woolwich be in prospect, he *must* go to one of the cramming schools to prepare for *that* examination.

Thanks for lending us Willy's letter: he is a delightful little fellow, and I have no doubt he will get on well everywhere. The notion of his being "head boy in Divinity" is indeed charming. Perhaps he will be a Bishop: though, indeed, who can tell whether thirty years hence a Bishop will have anything to live upon?

I am looking forward with very great pleasure to seeing dear Cissy and Emmy and George here next month. Georgey keeps up an active correspondence with Fanny. March is vindicating his character, and *making* up for the delightful mildness of February. We have a horrid shrivelling east wind, which makes me very cross, and disagrees very much with Fanny, and which will probably kill the buds. We are not so early here as you are, though this has been (to the end of February) the earliest season I can remember.

My cordial love to Cissy and your children.

Ever your affectionate Brother,

CHARLES J. F. BUNBURY.

I must write to you another day about politics.

JOURNAL.

We examined and discussed plans for new ser- 1869. vants' rooms. Yesterday, in *The Times*, saw the death of Sir John Boileau. He had long been in a wretched state of health; the last time I saw him, last June I think it was,—I thought he would hardly live till the winter. He was always very courteous and kindly to us, and we spent some pleasant days with him at Ketteringham, and met some agreeable company at his house in London. He was a man of much cultivation and refinement of mind, of considerable variety of knowledge.

We dined at the Arthur Herveys, just returned from London. Met the Maharajah, Lord and Lady Bristol, Lady Rayleigh, Miss Strutt, &c.

Accounts in the newspapers of a smart shock of earthquake felt in Lancashire and in the west of Yorkshire. No mischief done however. We dined early and went to the Athenæum to hear Lord John Hervey lecture on India—good.

Fanny went over to Mildenhall.

We chose places for planting some new trees. A

1869. number of new Ferns arrived, seut by Dr. Hooker. Lord and Lady Arthur and Sarah and Kate Hervey spent part of the afternoon with us.

Tuesday, March 23rd.

Spent all the morning at the Quarter Sessions at Bury—large attendance of magistrates—discussing about the police.

Wednesday, March 24th.

The result of the division on the second reading of the Irish Church Bill is a majority of 118 for the Government, in a very full house; a great triumph for them.

I received Wallace's new book on the Malay Archipelago, and began to read it.

Visit from Col.* and Mrs. Ward. Sarah and Kate Hervey arrived; we had a very pleasant evening of cheerful conversation with them.

Thursday, March 25th.

We had a very pleasant walk with dear Sarah and Kate Hervey about the grounds, and planted two trees for them—they went away after luncheon.

Sunday, March 28th.

Extraordinary weather. The 25th and 26th, especially the latter, were milder than the weather for some time before, the wind having changed

* Poor Law Inspector.

to the west. But when we rose yesterday, the ground was thickly and uniformly covered with *snow*, which had fallen in the night; it lay till the middle of the day, and though it melted in the afternoon, there was a fresh fall towards night ; this morning again, the coat of snow looked dazzling and unbroken enough for a traditional Christmas.

We went to morning Church and received the Sacrament.

LETTER.

Barton,
March 29th, 1869.

My Dear Henry,

I have been much interested and gratified by your letter of the 19th, and very glad to find that our thoughts as to dear Henry's education and prospects agree so well. I think you are quite right in this (and I had not seen it so clearly before) that the great public schools, with all their advantages as to moral and social training, have this disadvantage: their almost exclusive classical teaching is almost useless—is in fact, almost time thrown away—for boys who have to pass examinations for any of the active professions. I see in the papers that the head master of Harrow wishes to remedy this, and proposes to establish in his school a modern department ; in which modern languages and other branches of modern learning may be substituted for Greek and for the extreme study of Latin. If this be followed by the other great schools, and well carried out, there might be some hope of dispensing

1869. with the mere *cramming* schools.   But of course it
would be some time before it could be got into
working order.

I look forward with great delight to having dear
Cissy here, and talking over many matters with
her at leisure.   I long to see Emmy too.

I suppose the Irish Church establishment is
doomed.   The majority is tremendous, and though
I dare say the opposition will fight to the utmost, it
hardly seems to be a sort of measure that can be
beaten in detail in committee.

Assuming that the thing is to be done, I suppose
it is as well done by Gladstone's Bill as can be;
though indeed one ought not to speak confidently
about this without much more particular knowledge
of Ireland than I possess.

As to the general principle, I still say that I
should much have preferred to see all the three
Churches endowed, but they say this had become
impossible.   I dislike the voluntary system.

As to the reductions in the civil and military
services, the tide of popular feeling is running so
strongly that way, that I expect the reductions will
be carried much too far for efficiency, and after some
years there will be a re-action, and much of what is
now done will be undone.   I hope not too late.

With much love to dear Cissy, Emmy and the
boys.   Believe me ever,

Your very affectionate brother,

CHARLES J. F. BUNBURY.

LETTER.

Barton,

29th March, 1869.

My dear Lyell,

I cannot help thinking it a pity that Mr. 1869. Wallace, in his excellent book on the Malay Archipelago, did not get some friend to correct his botany. At page 184, he gives a list of what he calls *European* genera of plants found on the high mountains of Java; and from the context it is evident that he means genera which are specially and characteristically European. Now nearly half of those in his list are not so at all, but either groups of very wide distribution or such as merely struggle into Europe. Thus, Impatiens has *ninety-six* Indian species and one European; Lobelia, twenty-six in South Africa, a great many in Mexico, *two or three* in Europe; Oxalis, above one hundred in South Africa, and I believe one strictly European; Rhododendron, four in Europe and eight or ten times as many in India. Even Quercus and Vaccinium, according to Joseph Hooker, have their head-quarters *not* in Europe, but in sub-tropical Asia.

Again, Hypericum occurs in almost every country which is not very cold; and Polygonum is still more ubiquitous. I rather wonder that so great a naturalist as Mr. Wallace should have fallen into the common-place error of supposing that a group is specially European, because its European represen-

c

1869. tatives happen to be the most familiarly known. I am sure he would not have done so in the case of a zoological group.

I am very much interested in the book, particularly in the development of that very curious observation which he first published in the Linnean Journal, as to the two distinct zoological provinces in that region, and their boundary line in the narrow strait of Lomboll.

This weather is favourable for reading, as it by no means invites to out-door sauntering.   March seems inclined to march out in a very un-lamb-like fashion. Yet the buds of the trees are swelling, and the red American Maple, *Acer rubrum* is in full flower.

With much love to Mary,

Believe me ever, yours affectionately,
CHARLES J. F. BUNBURY.

JOURNAL.

Tuesday, March 30th.

We went to the Athenæum at Bury, and heard a lecture from Mr. Churchill Babington on the "Archæology of the New Testament," that is an explanation of the coins, weights and measures, implements of domestic use, and various other matters in use in Judea at that time, and mentioned in the New Testament.   It was full of good sound information given in a clear and unpretending manner.

Met the Arthur Herveys and Lady Cullum.

LETTER.

Barton,
April 1st, 1869.

My Dear Susan,

I owe you and dear Joanna many thanks 1869. (for I believe I have not yet thanked you for the pretty photographs which you sent me) and much more for the kind feeling which prompted you to send them. Since I wrote to Joanna, that is throughout last month, our course of life has been singularly quiet and uniform and uneventful; we have not even dined out, except once with the Arthur Herveys, and have scarcely seen anyone except at two lectures at the Bury Athenæum, to which we went. The most agreeable deviation from this uniformity has been that dear Sarah and Kate Hervey spent one evening and the next morning with us. They are truly charming girls; I can hardly imagine anything more delightful, indeed they two and Rose and Mary Kingsley, appear to me, patterns of everything that girls ought to be. You know me too well, I hope, to suppose that I have found this quiet time dull, or been inclined to wish for more society; it has been a very enjoyable time, especially since Fanny has got into better health (for she was not at all well in the early part of the month) and the days have passed only too quickly. I leave her to give an account of her own occupations. I am reading to her the "Life of Lord Brougham," and had before read that of " Lord

1869. Lyndhurst " in the posthumous volume of Lord
Campbell's " Lives of the Chancellors," edited by
his daughter.   They are very entertaining, but
written in a rather satirical spirit, and the book will
cause, I should think, a good deal of dissatisfaction,
and excite many attacks, for Lord Campbell has
something sharp to say of almost everyone whom he
mentions.

I am reading an entertaining old book, Burnet's
" History of His Own Time," more properly
memoirs than history ; to be taken, of course, with
many grains of allowance, for Burnet was very
evidently prejudiced and credulous, as well as very
vain ; but I believe he was honest in intention, and
his frank, diffuse, gossiping narrative, gives a very
lively idea of the times.   His style is excessively
careless, and his *pronouns (he, him, his,* &c.) sometimes get into a comical state of entanglement.
Also I am reading a spick and span new book,
Wallace's " Malay Archipelago," which interests
me very much, and would, I am sure, interest
Joanna ; it belongs to a class of books which I am
especially fond of, natural history travels, and is
remarkably well done ; it refers too, to a part of the
world rarely visited by Englishmen.

(April 2nd).   I have just received Joanna's very
kind and agreeable letter, dated March 26th, for
which pray give her my best thanks.

The pretty flower she enclosed (Gagea arvensis)
is not English and has no English name.   One
species of Gagea is found in England, but is very
rare.   The genus was named after Sir Thomas

Gage,* and is next of kin to the Star of Bethlehem. 1869.
I am extremely glad to hear you are both so happy,
spending your lives so pleasantly, and seeing so
many pleasant people.

Pray remember me kindly to Professor and
Madame Parlatore.

I always continue to occupy myself a little with
botany, and am making a careful study of Schimper's
" Synopsis of European Mosses." with a view to
understand thoroughly, his principles of arrangement.

Dr. Hooker letely sent us from Kew, living plants
of above forty kinds of ferns; a great acquisition
to our collection.

To conclude —we have nothing to complain of
except that we are in danger of being poisoned in
our house by *dead rats*

With much love to Joanna, believe me,

Your very affectionate brother,

CHARLES J. F. BUNBURY.

JOURNAL.

Thursday, April 1st.

We went out most of the afternoon. inspecting
the plantations and choosing places for new trees.

Monday, April 5th.

I went to Stowmarket and attended a meeting of

* Father of the last Sir Thomas and Sir Edward Gage·

1869. a committee of Magistrates of the two divisions to consider about union of the county police under one head.

_____

Wrote to Mr. Nicholl about Clement, who left us this afternoon.

_____

Business.   Signed Berry and Ortner's leases.

_____

A delicious summer day.   The 25th anniversary of our happy engagement.   We went to morning Church, and spent the afternoon in the arboretum, lounging very pleasantly.

_____

Fanny went to Mildenhall, and returned to dinner.

_____

Another delicious day.   We went to the Shrub. Fanny riding, I walking ; saw the anemones and primroses in full beauty ; afterwards strolled about the grounds, out nearly all the afternoon.

_____

We went in the open carriage to Cockfield—saw Mr. and Mrs. Baldwin ; walked with them and Scott through Bull's Wood, and over part of my

farm. Then called on Mr. and Mrs. Churchill 1869.
Babington. Saw their pretty house. Little Harry
arrived.

LETTER.

Barton,
April 18th, 1869.

My Dear Lyell,

A paper by Mr. Carruthers on Sigillaria,
was lately referred to me by the Geological Society,
and I recommended it to be printed. He gives
reasons for believing Sigillaria to be Cryptogamous,
in which I quite agree with him.

Bentham has lately read to the Linnean Society,
a paper on Cassia, which (to judge from the short
abstract of it in the *Gardeners' Chronicle*) must be
very important, especially in his general views. He
appears for the first time distinctly to adopt the
doctrine of the origin of species by variation.

Sir George Grey in a letter, which Fanny has had
from him to-day, speaks of the dreadful conflagration
which has devastated an extensive district at the
Cape, and says he means to send you an account of
it, as a curious illustration of the way in which a
great forest, with all the animals it contained, may
be destroyed, and the whole natural productions of
a large district changed at once—I am much afraid
that European or ubiquitous weeds will take the
place of the destroyed plants.

With much love to Mary, believe me,

Yours affectionately,

CHARLES J. F. BUNBURY.

JOURNAL.

1869. The long story that Burnet tells about the birth
of the son of James the Second (intended to prove
that the child was spurious) is a signal instance of
his credulity.   It is only worthy of some antiquated
coterie of scandal-mongers in a small country town.

———

We went into Bury—shopped ; then to Ickworth,
and saw dear Lady Arthur.   Home late.

———

Dear Cissy, Emmy and George arrived in the
evening.

———

Mr. and Lady Susan and Miss Milbank, Mr.
Arthur Duncombe, the Arthur Herveys and Mr.
Tyrell dined with us.

———

Our dinner party—the Abrahams, Lord John
Hervey, the Wilsons, the Miss Waddingtons, the
James Bevans, Mr. B. Bevan, Mr. Tyrell, Syden-
ham Hervey.

———

Dear Cissy and her beautiful children went away.

We had a long conversation about the site of a new 1869. vine house.

———

We went to morning Church and received the Communion.

Mr. Percy Smith came to luncheon.

———

Preparations for leaving home. We looked over some old papers of my grandfather Fox's, and arranged new books.

———

We arrived in London.

We dined with the Charles Lyells; met there Sir George Grey (the famous ex-Governor of New Zealand), Mr. Lecky (the author of the very interesting " History of Rationalism, &c.), and a few others,

Sir George Grey made upon me the same impression which he did when I first met him several years ago; that of a peculiarly grave, earnest, thoughtful man, of great strength of character and purpose. He expressed in the after-dinner talk, very democratic opinions, speaking strongly of the advantages resulting from universal suffrage in the Australian colonies (including New Zealand). He seemed to wish that the principal colonies of the British Empire should be independent, but connected with Britain by a sort of Federal union. He said that the most eminent men in the

1869. colonies were apt to be much disgusted when they came to England, at finding themselves looked upon as *provincials*, and perceiving that the quality of colonist is viewed as a token of inferiority.

In the evening there was a large party, including many whom I was glad to meet, and some of my particular friends.   Sarah and Kate Harvey, Minnie and Sarah Napier, Mrs. Evans Lombe and Miss Kinloch (a very pretty, interesting girl), George Napier and Bentham.

May 10th. Monday.

Went for a short time to the Royal Academy Exhibition, now open in its new apartments in Burlington House. The rooms are very beautiful; the pictures very well arranged, none placed too high or too low to be well seen.

Talk with Charles Lyell at the Athenæum. He thinks that there is an unnecessary panic about the probability of a war with America, though the temporary fall of American securities in the city was not owing (he says) to this fear, but mainly to narrower, and more purely commercial causes. He thinks Goldwin Smith in his alarm exaggerated. The letters of *his* correspondents from America indicate no apprehension at all of a rupture. Mr. Motley will come to this country he thinks, with a real desire to promote peace, and Grant will be ambitious to signalize his presidency by settling the disputed questions, and establishing peace on a firm basis. But it will be prudent to let the question lie dormant for the present, and not to attempt anything new in negociation; for the Americans are undoubtedly disgusted by Reverdy Johnson's unguarded speeches and by the apparent court he paid to some of those in this country who had been most conspicuous for their anti-American zeal. Lyell characterized Sumner's speech, as the speech of a madman.

The article in the April number of the *Quarterly Review* on Lyell's Geological Works was written by Wallace ("Malay Archipelago.")

Lyell says it goes beyond his views, both as to Glacialism and as to the development theory.

I went out in the carriage with Fanny, and we paid some visits; we saw Mr. Walrond, who was as charming as ever; also dear Rose Kingsley, who is staying at Mr. Froude's; also poor Mrs. Bowyer, in much anxiety about her husband, who is very ill.

Tuesday, 11th May.

Rose Kingsley came to us directly after luncheon and went out with us and Cissy to Mr. Theed's— afterwards to a party at Minnie's, where we met Norah Aberdare, Catty Napier, Pamela Miles, Blanche Napier, Lady Rayleigh; Edward dined with us.

May 13th, Thursday.

Went down to Folkstone.

The day very bright and fine with a very keen east wind. Even in travelling by railway I enjoyed the beauty of the rich and varied country of Kent,

1869. now in its fullest loveliness with the delicate colouring of the young leaves, the blossoms of the fruit trees, and the exquisite green of the grass fields. But the hop grounds showed nothing but bare poles.

––––––––

Folkstone.   Bitterly cold.

We took a long drive in an open carriage, through a delightful country, to Saltwood and Lynne, and returned by Hythe and along the shore. Saltwood Castle is a much greater and more picturesque one than my memory represented it. (I saw it in 1840). It stands well and covers a good deal of ground. The inner wall or enclosure is complete, and with its tower, has a dignified and imposing appearance; the moat between it and the outer wall is very broad and deep, its sides and bottom now covered with beautiful green turf affording a very agreeable walk and fine views. The gate-house in the inner circuit of walls remains entire, but being inhabited has modern windows put into it; its two flanking towers, round and very lofty, are said to be a century older than the gate-house itself. In the inner court there is no appearanee of a *keep*, but some remains of buildings particularly of what appears to have been a handsome banqueting hall.

Many fine trees, especially ash and maple, grow among the ruins, and on the sides of the moat; wall flowers in profuse blossom on the old walls which are haunted by abundance of jackdaws.

From Saltwood to Lynne we drove through 1869.
intricate and pleasant lanes, abounding with blue
bells and lychnis, primroses and ferns ; passing Mr.
Deede's beautiful park. Lynne Church is large,
but appears miserably neglected ; from the churchyard, where we were nearly blown away, the view is
striking. We were on the brink of a very steep, all
but precipitous, grassy declivity, descending to the
wide alluvial flat of Romney Marsh, which lay
spread out between us in all its extent, bounded by
the sea ; the view extending westward to the heights
near Hastings. On the green slope between us and
the Marsh, we could see the scattered masses of
Roman masonry of the Castrum Lemanis.

The excessively rough weather and want of time
prevented us from going down to examine it in
detail.

———

May 15th, Saturday.

We took a long drive into the chalk country to
the northward ; to Paddlesworth, Swinfield and
Acryse. First, a long ascent up the steep, bold face
of the chalk Downs, which rise very suddenly from
the lower grounds, and form a very striking continuous range of bold heights, with a well-marked
tendency to the conical shape in the prominent
portions. I agree with White of Selborne in finding
something very pleasant in the swells and undulations of these chalk hills, and the lights and shades
on their smooth grassy sides. As we looked down
from the upper part of the ascent, the lower country
between the Downs and the sea, in reality so varied,

1869. appeared a plain.   After reaching the top, there
was no descent for a distance of miles ; we went on
over a real table land, an open and nearly level
plain, much in pasture.   Further on, some pleasant
lanes and some steep descents into deep and pretty
valleys, but the general character of the country,
much more open and bare than the Green-sand
country we saw the day before.

Paddlesworth Church very small, very old with
Norman arches, far from beautiful.

Swinfield Church large—much neglected.   Remarkable round tower, side by side with the square
one, and looking as if it were attached to it, but
perhaps, in reality, older.

Acrise (what is the right spelling ?) Church, small,
recently renovated, and in beautiful order, a great
contrast to most of those we have seen hereabouts.
It is in the grounds of Acrise Court, now belonging
to Mr. Mackinnon (Lady Cullum's friend) a handsome, oldish, brick house, with magnificent trees
near it, and a noble rookery ; the park extensive, and with fine variety of ground.

————

Monday, 17th May.

In the afternoon we drove to Hythe and saw the
Church, which stands high in a commanding
situation, and has a very dignified appearance.
Interior handsome —very remarkable for the great
elevation of the chancel above the nave.   There is
an additional ascent from the chancel to where the
Communion table stands, so that the view from the
latter along the nave is very commanding.

Tuesday, May 18th.

Travelled from Folkstone to Canterbury. We took a first view of the Cathedral, and went rapidly through the interior, escorted by a little girl.

19th May, Wednesday.

We studied the Cathedral carefully in the morning, and called on the Dean (Alford). In the afternoon, the Dean took us through the library and other monastic buildings connected with the Cathedral. The library—fine. We dined with the Dean and Mrs. Alford—both very pleasant.

Thursday, 20th May.

Fine but cold. We went to St. Augustine's College. The warden was very courteous, and showed us the whole of it. Then we went to St. Martin's Church, the first founded in England.

In the afternoon we went to the curious old alms-house of Harbledown.

21st May, Friday.

In the morning Fanny attended a confirmation* in the Cathedral.

We took luncheon with the Dean, and met Archbishop Tait.

The Canterbury Cathedral has fully answered my expectations. Its beauty and majesty are such that

* The first confirmation Archbishop Tait held in that Cathedral,

1869. it is worth a long journey to see. I am doubtful
whether any other Cathedral that I have seen is
quite equal to it.

In the historical interest of its monuments, and of
the events connected with it, Canterbury surpasses
any other Cathedral I remember to have seen,
unless it be Westminster Abbey.

I was especially delighted to see the tomb of the
Black Prince, so well described by Stanley, whose
Memoir of that Prince (in his "Memorials of
Canterbury"), appears to me one of the most
beautiful historical sketches I have read. The
figure of the Prince on his tomb, "in his armour,
as he lived," a work of that age, and no doubt
representing his features and his armour with
perfect fidelity, is of extraordinary interest, and not
only so, but it appears to me a very excellent work
of art. Still more interesting are the Prince's
helmet, surcoat, shield and scabbard suspended
above the tomb.

The monument of King Henry IV. and his
Queen is also, historically, very remarkable, and
I looked with very great interest on the plain,
unadorned tomb of the great Stephen Langton;
and though I am far from having thorough un-
qualified veneration for Thomas Becket, it was
very interesting to see the actual spot where he was
murdered, for an abominable murder it undoubtedly was.

One of the finest general views of the Cathedral,
is from the top of the hill on the old London road
above the village and curious old hospital of

Harbledown, a mile or so fron the city gate.
Thence the whole majestic mass of the Cathedral,
and especially the great Tower, are seen rising
gloriously above the city, which clusters around it.

We left Canterbury by the express train about
4.30., and arrived at 48, Eaton Place, between 6
and 7, all safe, thank God.

May 22nd, Saturday.

Henry came to see us in the evening; just
arrived from Wales.

May 23rd, Sunday.

Called on Lady Smith* and Sir John Bell,† also
on Miss Phillips,—pleasant. Henry dined with us.

Wednesday, 26th May.

Visited the Royal Academy and met there the
Matthew Arnolds.

Friday, May 28th.

Went with Fanny to Gregory's in Regent Street
and ordered an ebony table. Charles and Mary,
Edward, Henry and George Napier dined with us.

30th May, Sunday.

The 25th anniversary of our happy wedding day.

---

* A Spanish lady, widow of Sir Harry Smith, who married her at the time of
the Peninsula War.—(F. J. B.)

† An old friend of Sir Charles, at the Cape in 1838.

1869. God be thanked for all His goodness to me, and
especially for the blessing of this marriage.

We went to morning Church. Visited Mrs.
Ellice. Minnie and Sarah lunched with us.

31st May, Monday.

Visited my poor old friend, Mr. George Jones,
the Royal Academician, now in his 84th year; found
him very feeble, but very kindly, cordial and cheerful.   He exhibits   three   pictures   in   this   year's
Exhibition, and very good ones; two of them
indeed — those of Cawnpore and Lucknow, were
painted before his illness of last year, having been
begun for Lord Clyde ; but that of the destruction of
Magdala has been done since his illness, and a very
remarkable work it is for a man of his great age.

Tuesday, June 1st.

Visit from Mrs. Phillimore, who was very agreeable ; also from Mr. Lott and Patrick Blake.

Our dinner party—Pamela and Phillip Miles,
Blanche Napier, Minnie, Helen Ellice, Kate
Hervey, George, Edward, Henry, etc.

Wednesday, June 2nd.

A visit from William Napier, just returned from
Spain; very glad to see him.

I went for a short time to the flower-show at the
Horticultural Gardens; Waterer's show of Rhododendrons, as usual, gloriously beautiful.

Thursday, June 3rd   1869.

We went (the two Hervey girls, Minnie and Sarah and little George) to the National Gallery, where we were met by Mr. Boxall, and he accompanied us through most of the new rooms (those lately occupied by the Royal Academy), showing us the new arrangement of the pictures. The improvement upon the old arrangement is very great, the pictures have room enough, they are not crowded, there is space enough for each to be placed in a favourable light and to be seen to the best advantage. We can now see, much better than before, how rich a collection it really is. I do not however perceive any very clear or definite principle of arrangement (as to schools and so-forth) and indeed Mr. Boxall seemed to think this a matter of no importance.

We dined with the Edward Romillys; went afterwards to Mary's party, which was very pleasant.

————

Saturday, June 5th.

Katharine showed me a collection of dried American Ferns, which she has lately received from Mr. Redfield, including a set, gathered and dried by himself, of characteristic Ferns of the old States, most beautifully preserved, and also many from California and New Mexico, mostly bad specimens but rare and curious.

A beautiful day.

Fanny went to Wimbledon, to the Athletic

1869. Sports, with little George. I went to see Katharine and she showed me some interesting specimens of Ferns.

We dined with Minnie, met Lady Lilford, Lady Rayleigh and her daughter, Mrs. Ellice, Captain Spencer, etc.

Sunday, June 6th.

It is reported that the Conservative majority in the House of Lords have determined to throw out the Irish Church Bill on the second reading. I am afraid this is very rash.

Monday, June 7th.

Henry set off for Wales.

I drove out with Fanny and Sarah Hervey, saw Lady Cullum and Mrs. Upton.

Wednesday, June 9th.

A delightful day. Went out driving with Fanny and Sarah Harvey—then to Norah Bruce's afternoon tea, met Mrs. Grey, Mrs. Phillimore and others.

A most pleasant quiet evening.

Thursday, June 10th.

Began reading a little of Dante with Fanny and Sarah. Mrs. Way came to luncheon. A visit from Miss Kennaway.

I looked at Minerals in the museum of practical Geology, and saw the Water-colour Exhibition.

A visit from MacMurdo. Two of the Ladies Legge came to luncheon with us.

The two Boileau girls, Sir John Kennaway, and Edward dined with us.

————

Saturday, June 12th.

We went, all three of us, to the wedding of Miss Millbank to Mr. Arthur Duncombe, at St. George's Hanover Square—a very brilliant affair.

Our dinner party—Mr. Powys and Lady Mary, the Bruces, the Louis Mallets, Miss Strutt and her brother John, Captain Spencer, Minnie, Mr. Lecky, George, Etc.

————

Sunday, June 13th.

We went to morning Church at St. Peter's.

Visit from Sir G. Grey. Interesting talk about New Zealand. He thinks the natives of that country the noblest uncivilized race with which we have ever come in contact. Yet they are convinced themselves, that they are dying out, that they are a perishing race; and many of them say that, as they know this to be so, they choose to die gloriously, fighting for their independence.

They have less fear of death than Europeans; they have no belief in future punishment, but believe that they shall certainly pass by death into a happier state. The mixed breeds, between them and the Europeans are very fine and handsome, and Sir George thinks it probable that the remains of the Maori race will gradually be absorbed by

1869. successive intermarriages and thus ultimately lost in the European.

The rejection of Christianity by the insurgents, and the setting up of a sort of hybrid religion in its place, no cause for despairing of the ultimate triumph of Christianity.

The Maoris now probably more civilized than the Anglo Saxons in the time of Augustine.

Sir G. Grey also told us that a learned man, who, supported by him, has long been studying the languages of Africa, has satisfied himself that the language of the "Bushmen" is entirely distinct from that of the Hottentots.

The Hottentot language, he says, is essentially of the same family as the Coptic. The Caffers, again, are a Negro race.

I will put down here a story I heard yesterday at dinner from Norah Bruce ;— While the proceedings against the Mayor of Cork were in agitation, before the Mayor had resigned, and while it was still expected that regular legal proceedings would be required, John Bright was talking on the subject with another of the Ministerial party, whose name I did not catch. Bright said, " I hear they mean to bring forward in the defence all the rash things I have ever said about Ireland." "Will the trial *ever* come to an end ?" said the other.

Another story (author Louis Mallet) but I heard it long ago—I think from Mr. Horner.— Sir Samuel Romilly invited Jeremy Bentham to dine quietly with him, adding " You will meet no one but George Wilson." Bentham's note in reply ran thus.—

" Dear Romilly, if we have nothing to say to each 1869. other, why dine ? if anything, why Wilson ?"

————

Tuesday, June 15th.

Spent much of the morning in reading the debate in the House of Lords on the Irish Church Bill: particularly interested by Lord Carnarvon's, the Archbishop of Canterbury's and Lord Stratford de Redcliffe's speeches.

In the afternoon, we two, with Kate Hervey, Mr. Hutchings, and little George, went to the Tower, and spent about two hours and a half in seeing it. Having an order from Lord de Ros,* we saw it thoroughly and at our leisure ; being shown the "reserved" parts, as well as those usually shown to tne public. It was very interesting, but we saw more, in the time, than it was easy to arrange or retain in one's head, and in particular I find it difficult to understand clearly the topography —the relative position of the different buildings ; but Lord de Ros's book and the articles (also very good) in Charles Knight's " London " give great help towards arranging and systematizing one's ideas on the subject. It is very interesting to see the actual cells in which famous prisoners were confined, the actual dungeons (and horrible dungeons some of them are), the inscriptions on some of the walls; the actual spot on the green within the court yard where the block stood on which Lady Jane Grey and so many others were beheaded —very interesting to view the scenes of so many

* Lord de Ros, Governor of the Tower at that time.

1869. horrible cruelties and sufferings. Especially the sight of the "Rat's Dungeon" and the "Little Ease," and the vault in which the torture used to be administered, bring vividly to one's mind the cruelty and wickedness which so terribly pervade our history as well as that of other nations.

Some of the interest attaching to the inscriptions made by prisoners, has been injured; by many of them having been cut out of the walls where they originally were, and inserted all together in the walls of one room of the Beauchamp tower.

A room in the Governor's house (now used as Lord de Ros's bedroom) is called the Council Chamber, and traditionally said to be the place where Guy Fawkes was examined by torture; but Lord de Ros seems doubtful of the truth of the tradition, and it is more probable that the torture was always administered in one of the more secret recesses of the prison; very likely in that gloomy vault at the bottom of the ———Tower from which the "Little Ease" dungeon is entered.

The armoury is also interesting, but in a different and less peculiar way. It is a fine and striking collection, but scarcely equal (as far as my memory serves me) to that in the Museé de l'Artillerie at Paris. The description of it in Knight's "London" seems to be very good.

Much of the armour of the 16th century and the latter part of the 15th is very beautifully and elaborately enriched, sometimes delicately inlaid with gold, sometimes engraved all over with a marvellous profusion of arabesques and fancy designs, and even of figures. I think some of the designs on these suits in the Tower are represented in the Archæologia.

The sight of these suits of armour, especially those of the 16th century, revived the wonder I have often felt, how any men could possibly move or breathe in such armour, or indeed how any horses could long support such weight. I cannot help suspecting that such excessively cumbrous and complete protection was worn on occasions of parade, and not in serious battle. Though indeed we read, that Knights once unhorsed lay totally helpless, and it often happened that some of them unwounded, were suffocated in their armour. The armour of Edward III.'s time, as represented on the Black Prince's tomb at Canterbury, looks much more convenient and serviceable.

I have been struck by a passage in " Montaigne" (book 2, essay 9) :—

"Il semble à la verité, a veoir le poids des nostres" (of our armour) "et leur espesseur, que nous ne cherchions qu'a nous deffendre, et en sommes plus chargez que couverts. Nous avons assez à faire à en soustenir le faix, entravez et contraincts comme si nous n'avions à combattre que de choc de nos armes."

Wednesday, June 16th.

Visit from Arthur Hervey—very glad to see him. We dined with the Youngs. Met the Bishops of

1869. Durham and Peterborough,* Sir Stafford and Lady and Miss Northcote, Mr. and Mrs Hodson.

Thursday, June 17th.

Mr. and Lady Mary Powys, and Cissy came to luncheon. Visit also from Katharine and the Kennaways. We dined with Sir John Hanmer.

Saturday, June 19th.

The second reading of the Irish Church Bill has been carried in the House of Lords by a majority of 33. It has been a splendid debate; a grand display of oratorical power; many first rate speeches.

Finding Lecky's book rather stiff reading for this gay time in London, I have laid it aside for the present, and began to read a new book by the Duc d'Aumale—"Histoire des Princes de Condé pendant les 16me and 17me Siecles." This promises to be very agreeable reading.

June 20th.

We dined with the Lombes.—Met the Adairs, Mr. De Grey and others.

Monday, June 21st.

This party at the Lombe's was very pleasant. Mr. De Grey† (the M. P. for West Norfolk) seems an interesting young man, quite an enthusiast for

---

The day after the Bishop of Peterborough made his remarkable speech on the Irish Church.—(F.J.B.)

† Now Lord Walsingham.

natural history, especially entomology, — an en- 1869.
enthusiasm which I like to see in a young man of
his station. He told me that the Struthiopteris
germanica, a Fern new to Britain, has been lately
found in a wood on his father's (Lord Walsingham's) estate in Norfolk. This is very interesting.
He has promised to send me fronds of it.

I have omitted to notice a very agreeable dinner
party at Mr. and Lady Mary Egerton's on the 18th.
Mr. Egerton appears a very agreeable man: and
there were besides, Sir Henry and Lady Rawlinson,
Sir Francis Doyle (the poet), Professor Tyndall,
Mr. Spottiswoode and Mr. Newton of the British
Museum.

———

Wednesday, June 23rd.

For some little time past, Fanny and dear Sarah
Hervey and I have been reading "Dante's
Purgatoria" together in the morning; and in spite
of frequent interruptions, we have gone steadily on
as far as the 6th canto. This morning we read the
grand passage of the meeting with Sordello, and
the magnificent outburst of indignation against the
state of Italy. "Ahi serva Italia, di dolore ostello,"
&c. &c.

It is delightful to read poetry with two such
women. I am reading the "Histoire des Princes
de Conde," and find it very entertaining.

I went to the Royal Academy, and spent some
time there. George Napier and Kate Hervey dined
with us. Fanny went to a ball with Sarah and
Kate.

1869.   Thursday. June 24th.

A dull, heavy, warm, yet not pleasant day. Our
dinner party—Lady Bell, Mrs. Grey, Mr. and Mrs.
Godfrey Lushington, the H. Lyells and Mrs.
Byrne, Mr. and Miss Donne, W. Napier, Mr.
Babbage, Cissy, Edward,—all pleasant.

———————

Friday, June 25th.

There is now exhibiting at the South Kensington
Museum a very fine collection of precious stones,
lent by Mr. Beresford Hope:—in particular, the
largest and most beautiful sapphire I ever saw, a *cats-
eye* of wonderful full-size and beauty ; another very
remarkable stone called also a *cats-eye*, (but query
whether really of the same nature ?), of a golden-
brown colour, with a lustre like that of waving silky
hair seen through water; a splendid beryl (said to
be the finest known) of a pure, though pale green,
not blue-green, and very transparent, formed into
the hilt of a sword, which belonged to Murat King
of Naples; a Mexican opal, very fine, the main
colour a greenish-grey instead of the milk-white of
the Hungarian opals.

A story told us by Norah Bruce:—

The Bishop of Peterborough, when he had
finished his famous speech in the House of Lords,
on the Irish Church Bill, was (very naturally) a
little flurried and excited, and as he sat down, he
took up the Bishop of Oxford's hat instead of his
own. The Bishop of Oxford said "I will exchange
heads with you, with the greatest pleasure."

———————

Saturday, June 26th.

Drove out with Fanny and Sarah. We visited
old Lady Lilford, who was most cordial and
pleasant. Met General Fox there. Then to the
Archbishop's at Lambeth, where there was a large
party.

———

Tuesday, June 29th.

Sir John and Miss Kennaway came to luncheon.
Our dinner party—Lady Rayleigh and Miss Strutt,
the Hanmers, Angersteins, Thornhills, etc.

———

Wednesday, June 30th.

Kate Hervey brought me several fine and fresh
specimens of the rare and very curious *Monotropa
hypopitys*, gathered at Ickworth, in the same grove
where she showed me last year a withered relic of
it. I never before saw it alive, and was much
pleased. I carried some of the specimens to
Katharine Lyell, to whom also it was new.

═══

LETTER.

48, Eaton Place, S.W.
July 1st, 1869.

My Dear Henry,

Very many thanks for your letter of the
25th, which interested me very much.

I assure you I have the greatest possible
pleasure in being able to help you and dearest
Cissy; there is no other application of the same
money which could have been so satisfactory to me,

1869. and I am truly glad that it has contributed
materially to your comfort.   I was sure that the
great amount of medical assistance required for
Cissy and Emmy, added to the education of your
boys, must be a heavy pull upon your resources ;
and I am truly happy to be able to help you.   I was
very sorry to say adieu to Cissy and Emmy ; it was
a great pleasure to have so much of their company,
and I hope they will have made out their journey
without catching any cold.   Emily is a dear, warm-
hearted girl, and we both love her very much.

I am much interested by your account of what
you have done and are doing at Abergwynant.   I
am sure you were very right to make it a great
object to build good cottages for your labourers, and
I do not doubt you will find yourself well repaid for
it.

Scott writes me word that the crops on my farm
and at Barton generally are now looking very well,
and I heard also from Mr. Thornhill, the other day
that there has lately been a very favourable change
in the prospects of the harvest, though it will
certainly be late.   They are very busy at Barton
with the hay, which is a remarkably abundant crop
(as it seems to be everywhere) and employs all the
labour that is to be had, though the mowing machine
(Scott writes) works very well.   This dry weather is
very favourable to the haymaking.

(July 2nd).   This consoles me for the uncomfortable coldness and *ungeniality* of the weather, which
is indeed very unlike June, but it has not spoilt the
roses at Great Barton, for those which are sent us

are very fine indeed ; Allen is remarkably success- 1869.
ful with them this year.

We have had a very pleasant time in London,
yet I shall not be sorry to find myself in our country
home again.

That most delightful and admirable girl —
Sarah Hervey leaves us, I believe, on the 13th,
and we shall probably go down on the 15th.

My best love to dearest Cissy and Emily and
your boys.

Ever your very affectionate brother,
CHARLES J. F. BUNBURY.

JOURNAL.

Thursday, July 1st.

Our dinner party—Mr. and Lady Mary Egerton,
young Lady Lilford, the Charles Lyells, Sir G.
Grey,* Sir G. Young, Minnie, etc.

Saturday, July 3rd.

We dined with Minnie. Met Mr. and Mrs.
Henry Hervey, Admiral Yelverton, Captain and
Mrs. Blundell and a few others.—In the evening
Miss Shirreff, Sally and her daughter, Miss Johnston,
etc.

Sunday, July 4th.

Fanny and I went down to the lovely place
Combe Hurst, and spent the afternoon with dear

* Governor of the Cape and New Zealand.

1869. old Mr. Sam Smith,* who was as agreeable and cordial as ever; we dined there.

———

Monday, July 5th.

We dined with the Angersteins, but Mrs. Angerstein unfortunately was ill and could not appear: a beautiful house and beautiful pictures.

———

Tuesday, July 6th.

We visited General Fox—very cordial and very pleasant. Mr. and Mrs. Mills and Edward dined with us—very pleasant.

———

Wednesday, July 7th.

I visited Norah Bruce where Fanny and Sarah joined me, and we went to a garden party at Holland House—beautiful grounds.

Minnie and Sarah came to us in the evening.

———

Thursday, July 8th.

We dined with the Gurdons: met Count Streleczki, Mr. Cox and Lady Wood, Sir F. and Lady Boileau and others.

———

Friday, July 9th.

Beautiful weather.

William Napier came to luncheon with us. He went with Fanny and Sarah and Kate to see the great cricket match at Lord's.

———

* The uncle of Miss Florence Nightingale—a very beautiful character.

(F. J. B.)

Went out with Fanny. Visited the Millses, and met Sarah and Kate and some of the Lady Legges there. We dined with Sir Henry and Lady Rich—a pleasant small party.

Count Streleczki (whom we met on Thursday evening at dinner at the Gurdons) said, that the natives of the Society Islands, Otaheite, &c., are so beautifully formed, that the finest Greek statues scarcely equal them in perfection of symmetry.

———

July 11th.

A very pleasant visit from Charles and Mary.

Here is a story told me yesterday by Lady Rich; When the Viceroy of Egypt was here, and attended a sitting at the House of Commons, he was struck with the noise the members made in calling out " Divide! Divide!"—and asked the gentleman who was in attendance on him, what it meant? The answer was, that they called out in that manner when they were tired of the debate and wished to put an end to it. Soon afterwards Sir John Bowring, visiting the Viceroy, became extremely prosy and long-winded in his discourse, when the Viceroy becoming intolerably weary of his prosing, at last broke out suddenly with "Divide! Divide!"

Another story. When the Archbishop of Dublin was speaking on the Irish Church Bill in the House of Lords, Lord Chelmsford and Lord Derby were sitting side by side listening to him. At last Lord Chelmsford said, "This is a very deep Trench."—

1869. "And a *very* long one," replied Lord Derby in a
doleful tone.

———

Visiting the Museum of Geology in Jermyn
Street, I saw with satisfaction that the important
collection of fossils of the Norfolk Forest-bed,
formed by Rev. Mr. King, has been deposited there.
It is rich, particularly in teeth of Elephas
meridionalis, and in cones of pine and fir, besides
many small seeds which cannot be identified
without minute examination. The fir cones are
easily recognized as those of the common spruce fir,
and many of them are in a state familiar to me
from my Mildenhall observations, having been
gnawed by squirrels, so that little is left but the
mere axis of the cone, with mere stumps of scales.
The pine cones have, I do not doubt, been rightly
identified as those of Pinus sylvestris, but they are
not exactly of the shape most common in that
species.

A very hot day.

Fanny and Sarah with a large party went to
spend the day at the Crystal Palace. I visited the
National Gallery, looked much at the Turners;
also visited the Museum of Practical Geology.

———

Visited with Fanny, Theresa and Lady Boileau
and Lady Gardiner. We dined quietly with the
Henry Lyells. We visited the Royal Academy
Exhibition—Fanny for the first time this year; I

have been there several times. I should hardly say
that it was one of the best Exhibitions I have ever
seen, but there are many good things; in particular,
three or four lovely landscapes by *Creswick*, and one
"Morning in the Meadows," by *Lee;* a most
pleasing cattle-piece, "An evening party," by
Cooper; a beautiful sea-shore view "On the Lagune
of Venice," by *Cooke; Landseer's* "Ptarmigan Hill,"
the dogs in this are admirable, and his two studies
of lions; *Millais'* charming portrait of a little girl
(Miss Lehmann);* a very clever and humorous
picture of common-life, "A disputed boundary," full
of spirit and character, by *E. Nicol.*

July 14th.

Wrote to Emily.

We all three went to the garden party at Holland
House—very brilliant, and really a very pretty sight.
Met many we knew.

Fanny and I dined with General Fox:—he was
very kind and pleasant: there were the Bruces and
others.

July 15th.

A very fine and hot day. Visit from Charles and
Mary and Katharine.

We dined early with Lady Rayleigh, but came
away early, when all the rest went to see fireworks
at the Crystal Palace.

---

* Now Lady Campbell.

We travelled down to Bury by the Sudbury line, in company with Sarah and Kate Hervey; parted with them (with much regret) at the Eastgate Station, and returned to our dear beloved home, where we found every thing in perfect order and comfort. Thanks be to God.

Saturday, July 17th.

At home. A splendid day. Arranging and settling ourselves at home: strolling about and enjoying the beauty of the flowers and trees, &c. Talk with Scott.

Sunday, July 18th.

My dear Wife's 55th birthday. A very hot day. We went to morning Church, and afterwards called on the Percy Smiths and saw their alterations at the Vicarage.

Monday, July 19th.

Walked through the groves to see our friends' trees. A visit from Lady Cullum. Edward arrived.

Tuesday, July 20th.

We went into Bury, got our tickets for the Archæological meeting—speeches—we met the dear Arthur Herveys. Mr. Clark of Trinity College, Cambridge, and Mr. Clark of Dowlais arrived.

A brilliant day.   We all went in two carriages to
Bury: saw the ruins and St. James' Church.   Walked
about the Abbey grounds a long time with Lady
Cullum—met Lady Adair and the Hugh Adairs.

———

Thursday, July 22nd.

A splendid day, very hot.   We went to a
luncheon party at the big house at Ickworth, a very
large gathering: afterwards went to the Arthur
Herveys and spent some time very agreeably with
them.   Mr. Gambier Parry arrived.

———

Friday, July 23rd.

Fine weather continuing.   Mr. Clark of Trinity
College went away.   Clara Strutt* and her brother
Charles arrived.   We all went to Hardwick, heard a
lecture from Professor Babington : met the Herveys,
etc.

———

Saturday, July 24th.

I showed the arboretum and garden to Mr.
Gambier Parry.   We all went to Hengrave to see
the house.   The Arthur Herveys dined with us
—also the J. Bevans and the Percy Smiths.

———

Sunday, July 25th.

Very hot: we all went to morning. Church and
afterwards went through the Church with Mr.
Gambier Parry.

———

* Now Mrs. John Paley.

Mr. Gambier Parry and Edward went away. Walked round my farm.

During the meeting of the Archæological Institute at Bury, this last week, we have had three very pleasant men staying with us:—Mr. Clark the Vice-Master of Trinity: *another* Mr. Clark of Dowlais in Glamorganshire, a neighbour and friend of the Henry Bruces, and Mr. Gambier Parry: the two latter were previously strangers to us. The Vice-Master I have repeatedly mentioned before, and he is a great favourite with both of us. Mr. Clark of Dowlais is very cheerful and good-humoured, a man who appears full of vitality and of vigour, both bodily and mental, well informed on a variety of subjects, having travelled much, observed much and read much.

Mr. Gambier Parry is a remarkably accomplished and agreeable man, of highly refined and cultivated taste, of great knowledge, especially in the fine arts and in mediæval antiquities, but by no means confined to those subjects. Without being a scientific botanist, he has much taste for exotic trees. I had great satisfaction in showing him the arboretum here, and found that his remarks showed much knowledge and appreciation of the subjects. He particularly admired our Pinus excelsa and Lambertiana, Abies Cephalonica, Catalpa and Æsculus indica, which last was (as may be supposed) new to him. He remarked that the Araucaria never thrives on a calcareous soil, and thought that its imperfect succes here was probably owing to too much calcareous matter mixed with our clays and loams. He had observed that the Douglas Fir seldom thrives much in this country after growing to the height of 50 or 60 feet. He has travelled in Dalmatia, a country seen by few civilized men, and he spoke of the wild grandeur of the scenery and the excessively savage and lawless habits of the people. He gave me, in particular, a description of a waterfall he saw in that country, where a broad river rushes in a succession of falls or rapids over a multitude of rocks of calcareous tufa, resembling gigantic sponges.

Here are two stories of school examinations, told by Mr. Clark of Dowlais. — At a school in *his* country (the iron district of Glamorganshire) something was said, in the course of an examination, about the Prophets. The question was put:— "What is meant by Prophets?"—Answer: "What the ironmasters make," (*profits.*)

A clergyman in the north of England asked one of his principal parishioners, a farmer, to examine his school a little in the catechism. The farmer, after a few other questions, asked—" What is your duty to your neighbour?" One of the children answered—"To believe in him." " Na, na," said the farmer, "that wunna do; if ye believe him, he'll *do* ye !"

LETTER.

Barton,
July 26th, 1869.

My Dear Henry,

1869.   I had some intention of writing to you
sooner after our return from London, but the
Archæological meeting at Bury has rather kept me
in a racket, or at least has rather distracted my
thoughts. Not but what we have taken the Archæology easily enough, keeping aloof from all the
excursions and most of the meetings; but we have
had company in the house, luncheons at Bury and
Ixworth, visits to Hardwick and so-forth.

The weather is splendid, and Barton is looking
very well; the grass very fresh and green, a great
contrast to its condition this time last year; on the
other hand, the foliage of the trees is not as rich and
luxuriant as it was last year, and in particular, the
plane trees are very poor in leaf, owing, I suppose,
to the frosts in May.   Pavia Indica in good flower;
the Catalpa with a great promise of blossom, but
none yet open; the white Lime the same.   What is
oddest is, that most of our young Wellingtonias, not
more than six or seven years planted, are bearing
well grown cones.   The fern-leaved beech is bearing
fruit, for the first time, as far as I know.

The wheat was looking still quite green when we
came down, but it has ripened very rapidly, and is
now of quite a different shade of colour; indeed

they say it is ripening rather too fast, and the crop, it is apprehended, is likely to be rather a light one on the light lands.

The farmers, I believe, would be glad of a little rain now, but there are as yet no serious complaints of the drought, as it seems there are in the south of Scotland.

Poor Alice Grey* is dead ; she seemed to be well when we came down from London, but soon after was attacked by a sort of paralysis, apparently without any assignable cause.

Give my very kindest love to dear Cissy and Emmy, and to your boys.

Ever your very affectionate Brother,
Charles J. F. Bunbury.

P.S. 1.—We heard a few days ago, from William Napier, *as a secret*, that Herbert has passed for Woolwich, which is very honorable to him, and very satisfactory to us as well as to his mother. Clement is with a reading party, somewhere on Loch Broom, in Rossshire. I hope he *will* read. You know how good a figure he made in the rifle contest at Wimbledon. His young idea has certainly learned to *shoot*.

P.S. 2. —I see in to-day's *Times* that Herbert not only has passed, but is 12th in a list of 40. It really is very good.

———

---

* A beautiful mare which once belonged to Colonel Bunbury.

JOURNAL.

1869.   Miss Strutt and her brother went away.

————

Mr. Burtt* came and spent the day in looking over our old parchments with Fanny. Scott tells me that the cutting of wheat ("Talavera" or white wheat) has begun at Mildenhall.

In this parish the wheat harvest is expected to begin next week. The ripening of the wheat has been very visibly rapid since we returned home twelve days ago.

Fanny lately received a letter from Lady Grey, who is staying in Norway with her husband, Sir Frederick, on a fishing excursion. She describes the place where they are staying as situated in a very deep and narrow valley;—"Here one is shut in "between two magnificent walls 4 or 5,000 feet "high, which leave only space for the river and a "strip of grass each side, and which cut us off from "the sun both early and late. His rays do not reach "us till 9.30, and are gone again by 3.30;—which is "an inestimable advantage to the fishermen, but to "them only." She says,— "Lilies of the Valley "grow in many places wild, and the oak and beech "Ferns are abundant and beautiful. The prin-"cipal wild flowers are, — Trientalis Europaea,

* Attached to the Public Record Office.

"Maianthemum bifolium, Menyanthes trifoliata, 1869.
" Linnaea borealis, Cornus Suecica, Polygonum (vivi-
" parum (?), Aconitum (Napellus) (?) Sedums and
" Saxifrages ; Potentillas and Vacciniums innume-
" rable. Plenty of the small Pedicularis, but not the
" *Sceptrum.*"

Resumed my reading of Lecky's " History of
European Morals," beginning with chap. 3.

LETTER.

Barton,
August 2nd, 1869.

My Dear Joanna,

1869.

Most of my time in London seems, as is often the case, to have been spent in a sort of busy idleness—pleasant enough at the time, but leaving little distinct impression on the memory. I did really however, find our stay in London very enjoyable. We had a good deal of pleasant society in a quietish way (decidedly quiet as far as I was concerned, for I went neither to balls, or late *drums*); we saw many old friends and made some agreeable new acquaintances ; and above all it was an immense pleasure to have that most charming and admirable of girls, Sarah Hervey, staying with us nearly all the time. I am fonder than ever of the Herveys, both Lord and Lady Arthur and their children. Nor were we entirely idle, for we read very carefully (*we*, including Sarah Hervey) through ten cantos of the "Purgatorio." Here Fanny was the leader, for I had read it only once before, and it was entirely new to Sarah. We also added to our knowledge in another way, by three very interesting days spent at Canterbury, with which we were both delighted— and by a very satisfactory day at the Tower, which strange to say I had never before seen.

I believe I told you in my last letter how much I had been delighted with Wallace's " Malay Archipelago " and, as I understand it has been sent to

you, I daresay you have read it with equal pleasure. 1869.
It is a book quite worthy to be classed with
Darwin's "Journal," and Bate's "Amazons" and
perhaps it contains more of novelty even than they.
In London I read with great pleasure the Duc
d'Aumale's "Histoire des Princes de Condé," of
which two volumes are as yet published; it is
written in a most agreeable style, and is both
entertaining and interesting in a high degree. The
greater part of the two volumes is occupied by the
Life of the first Prince de Condé, the famous
Huguenot leader in the wars of religion—a charming
character, though not without great faults. His son
and grandson were comparatively unimportant.
The next volumes I suppose will include the history
of the grand Condé.

I read also a Memoir of Harvey the botanist
whom I knew at the Cape. It is very pleasing and
gives, I have no doubt, a very true idea of his
thoroughly amiable and attractive character.

Lecky's book ("History of European Morals,
from Augustus to Charlemagne,") I began in
London, but found it a little too stiff reading under
the circumstances, therefore laid it aside for a time,
and have only now finished the first volume. The
first chapter is very long, very hard and rather dry,
and I do not as yet see that it is necessary to the
general scheme; it is an investigation of the
different theories that have been framed concerning
the origin of moral sentiments, with a zealous
advocacy of what is called the intuitive theory. The
remainder of the first volume is very interesting.

1869. Lecky's style is clear and agreeable (though perhaps not *quite* so good in this as in his first work), his moral tone always high and pure ; and he is especially remarkable for his extreme candour and fairness towards those with whom he disagrees.

I was much interested by your letter of May 25th though I have been so slow in answering it. I am very glad that you are going on with your book,* and trust in due time to see it published. I am sorry that the authorities at Florence are doing so much to spoil the antique beauty and interest of the town ; that mania for assimilating everything to Paris is very unfortunate, but I am glad that I have seen Florence *well* three times before its transformation. Certainly the tendency of modern *improvement* and civilization is to reduce everything to a uniform dull, tame, dead level.

I well remember the species of acacia you mention, called *Gaggia* by the Italians, and cultivated for its powerful and peculiar scent ; I remember it was a constant ingredient in the bouquets sold in the streets at Genoa ; it is the acacia (sometimes called *Vachellia*) *Farnesiana*, a native, I believe, of India. I have looked over the two numbers you sent me of the new Italian *Botanical Journal* by Signor Beccari and others ; it appears deserving of encouragement. I shall be very glad to subscribe (16s. per volume, as I understand) if you can tell me through what medium I can pay the money and receive the book.

* " Walks in Florence."

Our arboretum is looking well, but I do not think 1869.
the foliage of the trees is quite so rich this year as
last. It is odd that most of our young Wellingtonias only six or seven years planted, are bearing
cones.

The wheat harvest is just beginning about here,
last year it was quite finished before the end of
July.

I find that Fanny is writing to you, so I will
leave her to answer for herself. With much love
to Susan,

I am ever your very affectionate brother,
CHARLES J. F. BUNBURY.

JOURNAL.

Monday, August 2nd.

A visit from Professor and Mrs. Churchill Babington. We (2) dined at Ampton.* Met the
Judges of Assize, the Arthur Herveys, Lady Cullum,
the Hortons, the Wilsons.

Tuesday, August 3rd.

On the grand jury at the Assizes, all day. Lord
John Hervey foreman—a heavy calendar—dined
with the Judges, Ch. Justice Cockburn and Justice
Byles.

Wednesday, August 4th.

The Louis Mallets arrived.

* The Rodwells were living there.

Thursday, August 5th.

We all went to an afternoon party at the Arthur
Herveys ; some pleasant talk with dear Sarah and
Kate.

_____

                                    Monday, August 9th.

Mrs. Byrne went away.

_____

                                    Tuesday, August 10th.

The Arthur Herveys (with dear Sarah and Kate),
General and Mrs. Ellice, the Wilsons, and Mr.
Milbank, came to us in the afternoon, and walked
round the arboretum, and afterwards they (with
addition of Lady Cullum), dined with us.

_____

                                    Wednesday, August 11th.

We all dined with the James Bevans, I sat by
Mrs. Bevan, who was agreeable.

_____

                                    Thursday, August 12th.

The Louis Mallets went away.
An important note from the Archbishop to Fanny.
I answered it in part.

_____

                                    Saturday, August 14th.

We made various arrangements for leaving home.
Fanny went to Mildenhall in the afternoon, and
returned to dinner.

_____

                                    Monday, August 16th.

We left home at 7.15, and went up to London.
The William Napiers dined and spent the evening
with us. I called with William on old Lady
Campbell.

     * Pamela, Lady Campbell, daughter of Lord Edward Fitzgerald.

August 17th. 1869.

To Salisbury, by the 3.50 train from Waterloo Station, arriving at the White Hart, Salisbury, about 7. A pleasant journey. Striking contrast between the wild, interesting heath country of the Bagshot Sands (by Woking, Farnborough, &c.) with its extensive woods of fir and birch, and the bare, almost treeless, gently undulating, monotonous chalk country from Basingstoke to Salisbury.

————

August 18th.

Salisbury. We devoted three hours of the morning to seeing thoroughly the beautiful Cathedral. It is indeed singularly beautiful: characterized as specially (I should say) by beauty and grace, as Ely by solemn and commanding grandeur. It is best seen from the N. E. corner of the Close, and in that view the beauty of its general effects appears to me as great as any building can have.—Architecturally, it is remarkably interesting, as having been wholly built (except the spire and upper part of the tower) within 40 years, and therefore being all of one style—the early English in its highest perfection: hence a peculiar unity and simplicity of effect. The interior fine, but not of such striking and singular beauty as the exterior. The nave appears of only moderate height when we have the remembrance of Canterbury fresh in our minds. The clustered pillars (all the shafts, large as well as small, of Purbeck marble) very beautiful. Monuments numerous, and several of great interest.

1869. Two Earls of Salisbury, *both* bearing the name of William Long Espée: the first said to be son of Fair Rosamond and Henry II.: the other, *his* son, died in Egypt in the Crusade of Louis IX. Both recumbent figures in complete chain armour, interesting studies for the armour of the time, unfortunately a good deal injured. Sir John de Montacute, recumbent figure, in armour strikingly similar in style to that of the Black Prince at Canterbury, and of the same time he fought at Crecy.

Lord Hungerford of the time of Richard III.—recumbent figure, in alabaster, very rich and splendid plate armour, wrought in very elaborate detail.

Bishop Bridport (temp. Henry III.)—an extremely beautiful shrine-like tomb, with much sculpture, very finely executed, supposed to be by Italian artists.

A most remarkable *brass* of Bishop Wyvil (temp. Edward III.), who recovered Sherborne Castle for the See. The brass gives an elaborate bird's-eye view of the castle, with the Bishop blessing his champion.

Modern monuments.—The first Lord Malmesbury, by Chantry—fine. James Harris (the philosopher of Malmesbury) by Bacon.

Tablets to commemorate Bowles, the poet, and Maton, the naturalist.

The cloisters are of singular beauty; they reminded us much of the Campo Santo at Pisa.

There is something peculiarly pleasing and soothing in the look of the smooth bright-green turf, surrounded by those delicate arcades, with the 1869. magnificent Cathedral towering above. Nothing sad or gloomy in the scene. The late Bishop (Hamilton) is buried in the middle of the enclosure: the spot marked merely by a simple flat slab, with a cross and his initials.

The chapter-house is very beautiful and extremely remarkable, especially for its sculptures. The most curious of these are the series of heads, representing the head-dresses of various ages and conditions of life in those times.

After luncheon we left Salisbury by the 2.30 train, and went to Ottery Road Station, near Ottery St. Mary, where Sir John Kennaway's carriage met us. The day was beautiful, and the journey through a delightful country all the way, thoroughly English scenery, rich, varied, populous and smiling. Between Axminster and Ottery, especially, the country beautifully varied, steep and irregular, but fertile hills, little valleys with bright streams, much pasture land, luxuriant hedges and abundance of timber, small fields, many orchards.

———

Thursday, August 19th.

We went with the Kennaway party into Exeter, to the British Association Reception Rooms.—I to the section D:—lively discussion on protection of birds. Curious paper on Wheat. Met A. Newton, Joseph D. Hooker, &c.

———

1869.   Friday, August 20th.

We went into Exeter after luncheon ; went with
Sir John to the dog show, and afterwards to a large
garden party at Sir Stafford Northcote's,* a beautiful
place.

— — — —

Saturday, August 21st.

We two (Sir John Kennaway not being well) drove
into Ottery St. Mary's, and visited Sir John Coleridge
(the elder) who was very courteous, and showed us
the beautiful church, lately restored

Escot is three miles from Ottery St. Mary's, the
house is no way remarkable—it is quite new.

A very good portrait, full length, of Lord Corn-
wallis (the great one) to whom Sir John's father was
A. D. C.

There is also a portrait (believed to be by
Reynolds) of Major Lawrence, the first of our
Indian heroes. I do not remember to have before
seen a portrait of him.

The park is extensive and fine. The house
stands well on a slope, looking across a green valley
to the opposite slope (also in the park) which is well
diversified with abundance of fine trees, and
beyond it is seen a long high ridge of hill, bounding
on the E., the main valley of the Otter. Timber
very fine : almost all kinds of trees appear to grow
luxuriantly ; the beeches in particular superb, and
there are some grand old oaks, though it is said
they do not in general flourish so well on this soil as
the beech. Very near to the house stands one of

* We met Sir Bartle Frere there.—(F. J. B.)

the most picturesque and beautiful Scotch firs I 1869.
have ever seen ; a tree, I think, as beautiful as any
Italian pine.   In other places in the neighbourhood,
also, there are remarkably beautiful Scotch firs,
especially in Sir Stafford Northcote's park, and also
between Escot and Ottery.   In fact, an abundance
of fine and picturesque old trees of this species
appears to be a characteristic feature of this
neighbourhood,—one which I should not have
expected.

At Escot, a beautiful little flower-garden at the
end of the house, backed by a fine bank of wood.
Rhododendrons, azaleas, and all those things which
require a *heath* soil grow here most luxuriantly.   One
of the approaches to the house is by a long drive
along the side of a hill, through a perfect forest of
rhododendrons ; growing quite like wild shrubs—
just as they do in some places in Wales.

Hydrangeas also very fine, both with blue and
pink flowers.

The Evergreen Oaks here are of extraordinary
size and luxuriance.

There are delightful walks in the woods on the
slope above the house, and the luxuriance of the
vegetation of all kinds is very remarkable —semi-
tropical.   Many of these specimens of exotic trees
are very fine ; in particular, a grand cork tree, most
vigorous and picturesque, with just the same wild
freedom and irregularity of growth which one sees
in the South ; and the finest Liquid-ambar I ever
saw, really a tall timber tree.

On the 19th we went into Exeter, and I attended

1869. section D of the British Association.   Heard a
lively and rather entertaining discussion on the
protection of wild birds; conducted by Mr. Trist-
ram, Miss Becker, Alfred Newton, Sir John
Lubbock and Mr. Wallace, on the one side, and
Huxley alone on the other.   Also a very curious
paper on the improvement of wheat and other
grains by methodical *selection.*   Much the most
interesting person with whom we have made
acquaintance during these days is Sir John
Coleridge, the Solicitor-general, who was our companion in a railway carriage, to and fro, between
Escot and Exeter.   He made a very strong and
agreeable impression on me, and I should be very
glad to see more of him.   We visited his father, the
old Sir John Coleridge, by appointment of Sir John
Kennaway, at his house in Ottery St. Mary, on the
21st.   He was very courteous, and showed us the
Church of Ottery, which is a very fine as well as
curious one, and has lately been beautifully restored
mainly through his exertions.   Here we again met
for a few minutes the Solicitor-general.

August 23rd.

We left Escot.   Went into Exeter to section E
(Geography).   Then with Sir Jonn Kennaway, to
the Cathedral, but saw it imperfectly.

We left Exeter by the North Devon railway, soon
after 4, and reached Bideford about 6.15, drove
thence some four or five miles, to the Westward Ho !
Hotel, on the shore of the Bristol Channel, adjoin-
ing Northam Burrows and the famous pebble ridge.

The line from Exeter to Barnstaple through a 1869.
pleasant, rich, cheerful, varied country, with much
wood, the latter part along the valley of the Taw,
very pretty. From Barnstaple to Bideford,
beautiful views across the broad estuary, now a fine
glassy sheet of water, enlivened by many sails. In
the evening we walked a little way along the pebbled
ridge, which by its steepness and the size of its
pebbles, reminded us of the beach near Santa Cruz
in Teneriffe.

————

August 24th.

At the Westward Ho! Hotel, the weather delightful, as it has been ever since we left London.

We drove in an open carriage to Clovelly, setting
out about 11, and returning about 5.45. The
distance, twelve miles; of which the last three
before reaching Clovelly are through the woods
belonging to the Fanes. This three miles drive
is delightful, winding along the seaward faces of
excessively steep hills, through most beautiful and
luxuriant woods, which everywhere clothe these hills
or cliffs almost to the water edge; winding in-and-
out, across deep dells, which run down to the shore,
and are filled with luxuriant vegetation; at every
step most lovely peeps of the bright, blue sea,
opening between the trees. The heath most
luxuriant in these woods, and the banks mantled with
ferns and mosses in profusion; the Clovelly woods are
mainly of oak, sessiliflora, I believe. Holly in great
abundance and of remarkable size and beauty.
Amidst the herbaceous under-growth, the great

1869. wood-rush, Luzula sylvatica, is very conspicuous by its size and abundance.

Clovelly. This little town is as extraordinary in situation, and as picturesque in appearance as my memory represented it. Quite unlike an English town or village ; much more reminding one of a little fishing town on the Mediterranean coasts, or in Madeira.

The excessive steepness of the street, with its slippery pebble pavement, is very like Funchal ; the brightly white-washed slate-roofed cottages, with terraces before them, and luxuriant fuchsias beside the doors, have also an outlandish appearance.

The view from below is extremely pretty ; the white houses are seen rising steeply one above another, half embosomed in the foliage. The harbour and pier are exactly those of *Aberalva*, in " Two Years Ago."

* * *

August 25th.

We remained quiet at the Westward Ho! Hotel. The Pebble Ridge is very curious. It extends nearly two miles from south to north, and is especially remarkable for its great regularity and definitness ; the line between its base and the sands of the sea-shore, on the one side, and between its base and the turf-covered sands of Northam Burrows, on the other, is singularly distinct. The seaward (westward) face of the ridge, which is the steeper, has not a uniform slope, but is interrupted by two narrow horizontal terraces, where the materials are the same, but the slope very much 1869. less; and these lines of terrace run for a great length with striking regularity.

The eastward face has a continuous slope. The pebbles are very large, and their average size appeared to me about the same the whole way. At its northern end the ridge loses itself in a wilderness of high sand hills, steep and strangely broken, which border the mouth of the estuary. These are covered as usual with the Arundo arenaria, among which Euphorbia paralias grows in great abundance and vigour; but I could find no other plants.

The low and very level plain (probably of very recent formation) called Northam Burrows, bounded on the west by the pebble ridge—consists of sand mixed more or less in different parts with black peaty soil, and is covered with a very close and minute vegetation, in which I saw nothing peculiar. Here and there large clumps of Juncus maritimus. Much of this close herbage consists of Potentilla anserina, Plantago maritima, and Plantago Coronopus, all in a very diminutive state.

I find the Pebble Ridge described in a paper by Sedgwick and Murchison, in the Geological Transactions.* They observe that at first it seemed difficult to account for the origin of this remarkable ridge of gravel, but that it seemed to be explained by a *raised beach* which they discovered on the northern side of the bay "for we could not avoid "inferring that this Pebble Ridge was one of the "indications of a considerable rise of land." They

* Second series, vol 5, p. 282.

1869. think that it is not to be accounted for by the actual
configuration of the coast and the ordinary action of
the sea.

The Westward Ho! Hotel. This is a most
agreeable situation; the house stands just sufficiently
high above the shore, with an extensive and delightful view over the sea ; the open sea as it appears,
for the opposite coast of the channel is quite out of
sight; and in two minutes walking one is on the
beach. At high tide the sea washes the very base
of the Pebble Ridge ; at low water there is a great
extent of level and tolerably firm sand, with
scarcely a pebble. Behind the hotel and stretching
far away to the westward, is a steep, high bank or
face of hill (the edges in fact, of a table-land),
clothed with furze and heath. The shore below
(west of the Pebble Ridge) is bordered with low, but
very rough rocks of dark coloured slate.

LETTER.

"Westward Ho!" Hotel,
Bideford Bay, Devonshire,
August 25th, 1869.

My Dear Katharine,

Very many thanks for your letter of the
21st, which has given me great pleasure. I am
delighted to hear that Murray has accepted your
book. I am sure it will be an interesting and
valuable book to all botanists, and I look forward
with great pleasure to consulting it frequently.

We arrived the day before yesterday at this

large new hotel, which stands on the shore of the 1869. Bristol Channel, four or five miles N. W. of Bideford, just at the beginning of a very curious and celebrated *Pebble Ridge*:—a natural mound or ridge, singularly well defined, and two miles long, composed entirely of huge pebbles, and keeping off the sea from a broad low grassy flat called Northern Burrows.

The weather is delicious, and I am mightily enjoying the air and the beautiful view of the sea. Yesterday we made an excursion to Clovelly, a most curious place, approached by a drive of three miles through most lovely woods, which clothe the steep faces of the hills (one might almost say cliffs) down to the edge of the sea.

We spent four day's at Escot, a beautiful place, and the Kennaways were very kind, but we did not see much of the Association. By far the most interesting person we made acquaintance with, was the Solicitor General, Sir John Coleridge, who made a great impression on me, and of whom I should much like to see more. We visited his father, the ex-Judge, at Ottery St. Mary, and he showed us the fine and curious old Church, which has been beautifully restored. We propose to go to-morrow to Ilfracombe, where we have secured rooms at the Hotel.

Ever your affectionate brother,

CHARLES J. F. BUNBURY.

JOURNAL.

1869.   Still the same splendid weather, extremely hot.
We travelled in an open carriage from Westward
Ho! to Ilfracombe, resting at Barnstaple for about
four hours in the heat of the day.

At Bideford, we particularly noticed the very
long bridge of a surprising number of arches, very
old, with a new parapet: also the pretty situation of
the town.*

Barnstaple seems a consderable and thriving
town, with handsome streets and good shops, and a
look of traffic and activity.   Thence, a pleasant
drive to Ilfracombe, up a very pretty winding valley,
between steep hills: the varied grouping of which
as seen from the different turns of the road, is very
agreeable to the eye ; rich banks of woods, bright-
green meadows in the bottoms, the upland pastures
much burnt.   Part of this ascent is through the
parish of Marwood ; classical in geological story, as
one of the localites where Sedgwick and Murchison
studied the *Devonian* series of rocks, and collected
fossils.   The *Marwood beds* are upper Devonian.
Afterwards over a high bare open table-land, all,
however, enclosed and claimed.   Finally a long
descent to Ilfracombe.

---

* Beautifully described in the beginning of Kingsley's " Westward Ho!"

At Ilfracombe.

The strikingly wild and remarkable forms of the rocks here, ("shark's-tooth rocks," as Kingsley calls them,) the bold headlands and beautiful inlets of the sea, are just as I remembered them, and as I noted in my Journal of '41. But Ilfracombe, itself, has grown into a gay and bustling town, swarming with smartly dressed excursionists.

In the afternoon we drove to Watermouth, and a pretty little cove just beyond it where the Adiantum is said to have grown, but it seems to have been eradicated by collectors. Here (the weather being deliciously fine and calm) we took a boat, and were rowed a little way along the shore, enjoying the beauty of the sea and admiring the bold forms of the rocks and the dark caverns hollowed in them by the waves.

Crithmum maritimum in great abundance on the slaty cliffs about Ilfracombe, but everywhere out of my reach.

The satiny lustre on the faces of the laminæ of the slates is striking: also the great abundance of quartz veins which intersect them, and which often give them a curiously variegated appearance.

The great headland called Hangman Hill seen from the cove near Watermouth, appeared remarkably red in the distance.

The *stratification*, as distinguished from the *lamination* of the schistose rocks, is very distinctly seen in some of the rocks at Ilfracombe. There is

1869. an error here: what I have called lamination is real slaty cleavage.*

The rocks, it appears, are traversed by three distinct sets of divisions: bedding, joints and slaty cleavage.

———————

August 28th.

An extremely hot day.

From Ilfracombe to Lynton, a tedious journey of nearly four hours: road very hilly, dust excessive, country bare (though enclosed), monotonous and dreary, till we entered the valley of the West Lyn and began to descend towards Lynton, between steep wooded hills.

———————

August 29th.

Lynton. (The Valley of Rocks Hotel—excellently well situated, and fairly comfortable.) A sudden change of weather—a very violent and cold northerly wind and heavy threatening clouds, but no rain came. We went to morning Church, and in the afternoon drove in an open carriage to see some of the principal views. There is now a carriage-road through the Valley of Rocks and easy paths up to the principal crags, but the effect of those beautiful and interesting rocks is not at all injured, and I recognized with pleasure the spots which were so familiar to me 38 years ago. The violence of the wind, however, prevented us from rambling much.

*See Sedgwick and Murchison on the Geology of Devonshire,—Geol. Trans., ser. 2, v. 5, p. 646; where the very complicated structure of these Ilfracombe rocks is most carefully described.

Through the valley of the East Lyn, also, there 1869.
is now a good road, but here it is a disadvantage:
one does not half enjoy the scenery as one used to
do when one followed the footpath by the side of
the stream. Still it is a very lovely valley, and I
am happy to see that its beautiful woods have not
been diminished. The woods in the valley of the
East Lyn are mainly of oak, with some ash. The
oak is of the sub-species sessiliflora,—at least,
judging by its leaves.

                                                        August 30th.

Travelled in an open carriage to Dunster, twenty-
two miles: (nearly four hours), an interesting
journey. First, a very long ascent from the valley
of the Lyn to Countisbury, then many miles over
fine bold heathy hills of very considerable height
(part of Exmoor), overlooking the sea. A long and
steep descent to Porlock, a picturesque village or
little town, in a fine fertile vale near the sea ; a
grand old yew tree in the churchyard : myrtles
growing luxuriantly in the open air against the
fronts of the cottages. Thence through a rich and
varied country to Dunster.

                                                        August 31st.

Dunster. The Luttrell Arms, a very old house,
with a most remarkable sculptured chimney-piece in
one of the rooms ; situation excellent.

This is a quaint, picturesque, quiet little town,
consisting mainly of one street ; rising at one end to

1869. a richly wooded hill on which stands the castle; at the other to a higher and very conspicuous, almost conical hill, also clothed with wood and topped by a single tower. We went up to the castle, and saw the grounds, not remarkable except for the extreme beauty of the views from them. Indeed the scenery all about Dunster is of the most delightful kind; the most admirable mixture of hill and dale; of the richest and most smiling cultivated vales, slopes clothed with luxuriant wood, high, bold hills covered with heath and furze; and bounding the whole, on the one side, the sea, and on the other, the bleak heights of Exmoor.

Dunster.

A very remarkable and picturesque market cross, commonly called the Yarnbarn, situated in the middle of the street, opposite the inn.

When I speak of the *grounds* of Dunster Castle as not remarkable, I mean those immediately about the castle. The park is of great extent, and finely varied in surface, and appears to be well worth exploring.

A yew hedge, near the castle, is very remarkable, by far the tallest and largest I have ever seen. Grabhurst or Grabbist hill, near Minehead—seemingly an outlier of Exmoor—a very conspicuous feature in the view; a high, bold ridge, steep, but of rounded outlines; the flanks cultivated, the upper part clothed with furze and heath.

We had a very pleasant drive in the afternoon through the village of Wotton Courtney, where there is a grand old yew tree in the Churchyard, to Minehead

and down to the sea side. At Minehead, noticed
the luxuriant growth of myrtles (now in full bloom)
against the walls of the houses, and rising quite up
to the eaves; passion flowers also in blossom in the
open air.

Grabbist hill is stated by Mr. Horner to be 906
feet above low water mark.

" The highest part of the ridge is immediately
"above the village of Wotton Courtney." (Mr.
Horner).

Grabbist hill, as well as the Quantocks and
Exmoor, and all the intermediate hills are referred
by the most modern geologists, to the Old Red or
Devonian.

———— ————

September 1st.

Dunster (by road) to Bridgewater, through a very
pretty country; hilly, rich, and varied; skirting and
partly rounding the Quantock hills which are
bold and massy, though not rocky.

Great abundance of orchards, now in beauty with
their richly-coloured fruit. The soil deep red
throughout this tract. Many quarries of a hard
dark red or chocolate coloured stone, which
appeared to be a very compact sandstone or small
grained conglomerate.

We stopped to see the ruins of Cleve Abbey.
they are occupied as farm buildings, and are in a
wretched state of neglect; the principal part
remaining is a refectory of splendid dimensions,
with a finely arched and sculptured wooden roof,
and very large windows.

1869.   Outside, near the gateway, is an immense
sycamore tree, by far the largest I have ever seen ;
its huge limbs, spreading out near the ground, are
themselves like great trees.

We stopped for a little while at Nether Stowey,
celebrated as the residence of Coleridge, but not
otherwise remarkable.   It is at some distance from
the higher parts of the Quantocks.

Bridgewater.   An odious noisy town.

————

September 2nd.

From Bridgewater by railway to Glastonbury.
The country in general a dead flat, but between
Bridgewater and the Highbridge station we pass
through one hill* by a cutting, in which the stratification of the blue lias is well shown.   From Highbridge station to Glastonbury, a continuous extent
of absolute fen as level as those of Cambridgeshire,
and more nearly in their primitive condition, only
partially reclaimed ; much peat cut and piled up in
large stacks.   The Mendip hills and their outliers
visible in the distance.

Glastonbury :— The ruins of the Abbey very
beautiful ; they are inclosed in the grounds of a Mr.
Austin, and are very well taken care of, and very
liberally shown.   The principal remains are those
of the Church, which must have been a truly noble
one ;   the two tower piers are of magnificent height.
The best preserved portion is what is called St.
Joseph's Chapel, which seems to have been a sort
of appendage to the west end of the nave ;. this

* A part of the Polden hills.

(except the roof) is well preserved, and it is very 1869.
beautiful. In another enclosure, hard by, is the
Abbot's kitchen, a very singular structure.

Glastonbury Tor, a bold hill, rising into a remarkable cone, immediately behind the town, and
crowned by an old tower, is a conspicuous land
mark for all the country far and wide.

The inn, the George and White Hart, is itself a
very old house, and its front covered with very
remarkable sculpture, of the time of Edward IV.

From thence a pleasant drive of six miles to
Wells, through a very pretty country; the Mendip
hills full in front, and many bold hills on various
sides, some like islands rising out of the plain,
others projecting into it as bold headlands.

————

September 3rd.

Wells. A charming place. A very clean, quiet,
well-built, small town, in the midst of a lovely
country. We spent the morning in seeing the
Cathedral, which is glorious. The great west
front particularly fine ; more remarkable for width
than height, but especially striking from the
profusion and beauty of its sculptures ; absolutely
covered with statues. The effect singularly rich
and noble. The two side-towers of that front strike
me as rather squat, perhaps the only drawback to
the beauty of the whole. Interior:—the nave has an
effect of simple beauty and grandeur on the whole,
but the richness of detail is surprising, especially in
the capitals of the pillars, where heads and whole

1869. figures full of character and expression, are mingled
in the most curious way with the foliage.   The
inverted arches under the tower are exceedingly
remarkable.   The choir, the " retrochoir," and the
lady chapel, are of extraordinary beauty.

The Bishop's Palace is very remarkable, enclosed
by strong walls and towers, and a moat, rendering it
a regular feudal castle.   The broad moat is still
kept up, well filled with water, fed from copious and
constant springs which break out in the Bishop's
garden, and from which also the town is supplied
with water.   The garden is beautiful.

In the afternoon, we drove over the Mendip hills,
by the village of Priddy, to Chedder, returning
te Wells along the foot of the hills.   The first
ascent from Wells brought us on to an extensive
table land, nearly flat, and very open and bare, with
stone fences and no trees : very like the uplands of
Derbyshire (as about Buxton), and of the same
rock formation.   From this plateau we gradually
descend into the gorge of Chedder ; a deep defile
or ravine, in fact a deep cleft in the great limestone mass, running down from the table land to
the plain ; of considerable length, winding and in
general so narrow as to leave room for no more than
the road.   The grey limestone rocks on each side
rising into very grand precipices and crags, like
walls and towers of gigantic size.   The scenery is
equal, as to the rocks, to any of the Derbyshire
dales, which it much resembles in character, but the
extreme aridity is a defect.   The rocks are almost
utterly bare of vegetation, and there is not a trace of

water, not even the dry bed of a torrent, till we 1869.
came down very near to the town or village of
Chedder. Probably the course of the stream is
subterraneous.

I found Polypodium calcareum growing amidst
the fallen fragments of limestone.

From Chedder, the drive back to Wells, skirting
the base of the Mendips, is very pretty. On our
right, a rich plain with many insulated knolls and
larger hills rising out of it. Glastonbury Tor conspicuous in the distance.

———

September 4th.

From Wells to Salisbury by railway, by Frome
and Westbury. After an early dinner at Salisbury,
we immediately set off in an open carriage to see
old Sarum and Stonehenge, and accomplished it
comfortably in between four and five hours,
allowing time for sketching. Old Sarum is a very
fine specimen of an ancient entrenchment, believed
to be British: reminding me much of the Here-
fordshire Beacon in the Malvern Hills. The citadel
must have been very strong in those times: the
earthworks, ditch and rampart are on a great scale,
but there remain only fragments of the masonry,
probably Roman. The total disappearance of the
town, which in the Saxon and even in the early
Norman time was inclosed within the outer en-
trenchments, is very strange;—a populous and
important town with a Cathedral. The transfer of
the te, indeed, is not strange : one easily under-

1869. stands it as one looks down from the arid chalk heights, over the fertile and well-watered meadow-basin in which lies Salisbury.

Stonehenge is very impressive : much more so than I expected. It is exactly as it has been so often described and represented: yet there is something strangely impressive in the *reality*,—in the actual sight of those huge stones so mysteriously set up in times of unknown antiquity: those who set them up, and the purpose for which they were set up, alike uncertain. The effect is much heightened by the loneliness of the situation: for though there is a farm house at no great distance, and one or two small plantations within sight, these interfere but little with the general character of the open, untilled, unbroken, smooth grassy plain, which extends in almost all directions to the horizon. Cultivation has as yet done very little to break or vary this grassy plain: the monotony of which has in itself something of solemnity.

The stones of Stonehenge are so entirely coated with lichens, that it is only where they have been purposely chipped that one can see anything of their texture.

The dwarf thistle, Carduus acaulis, is most remarkably abundant amidst the turf around Stonehenge.

Along the road by which we returned to Salisbury, the country retains this perfectly open character for many miles : though it is indeed not a perfect plain, but gently undulated.

Salisbury.

Spent the morning in seeing the Blackmore Museum, shown to me by Mr. Stevens, the curator, who very good-naturedly devoted nearly two hours to showing it to me in great detail, and expounding his views. He is evidently a very clever man, and a specially learned man in the subjects of Post-tertiary geology and Pre-historic antiquities, about which he is quite an enthusiast. The collection (formed almost entirely, as I understood him, within the last ten years) is a wonderfully rich one, and beautifully arranged. In the first place are the remains of Mammalia, found in the drift-beds near Salisbury, and mentioned in Lyell's "Principles" (last edition):—the Mammoth, Rhinoceros, Reindeer, Lemming, Spermophilus (or Greenland Marmot) and many others: also the egg-shells (supposed of the wild Goose). Next, a vast series of the rude flint implements of the "Palæolithic" age : many from the immediate neighbourhood of Salisbury, others from all the localities in Britain, where such things have been found, from the famous places in the valley of the Somme, and so forth. A good many specimens of the modern forgeries of these implements.—Also some which are natural, but have a deceptive likeness to the implements. Mr. Stevens showed me some of these which I should certainly have taken for "flint hatchets," but which he assured me were merely accidental imitations : and he explained to me fully, and showed on the specimens themselves, the characters

1869. by which he could with certainty distinguish the
stones shaped by man from those shaped by nature.
Then come the relics, more numerous and various,
of the *Neolithic* or later stone age. There is a very
great variety among the stone tools and weapons
referred to this age : and some are so rough that I
should certainly have confused them with the
Palæolithic : but Mr. Stevens assured me he could
always distinguish them. The collection of these
articles is immense, from many parts of the world,
and in various materials: a great many from the
caves in the south of France :—and there is also an
instructive series of weapons and tools of modern
savages, illustrating the antiquities. Every thing
admirably arranged and labelled.

I became satisfied that this branch of research,
new as it is, has already grown into a science in
itself, and that Mr. Stevens is most deeply versed
in it. Mr. Stevens told me that he was confident
he could on an examination of any worked flint,
name the place from whence it came : that is,
supposing it was derived from any already-known
locality. There are many different types of form
among even the rudest Palæolithic implements, and
it appears that almost every locality (at least every
district) has its characteristic type. A large
number of Palæolithic implements from the valleys
of the Lark and Ouse,—from Bury, Icklingham,
Mildenhall, Thetford and Brandon : many given by
Mr. Prigg.*

* A geologist at Bury St. Edmund's.

Generally, the real Palæolithic flint tools have a 1869. certain porcelain-like or glazed appearance on the surface, which is characteristic : but one that Mr. Stevens showed me was black on the worked surface, and so fresh-looking that I should have doubted its antiquity. He, however, showed me some very minute marks by which he was satisfied that it was genuine and antique.

I should not omit to mention a large collection of curious specimens from the Swiss Lake-dwellings— both works of art—and remains of plants and animals used by the natives. The different seeds and fruits are, many of them, in a very perfect condition and easily recognized. There are also numerous examples of the (so-called) Bronze age : but when we came to them my time was running short, and my attention almost fatigued.

Monday, September 6th,

We went by railway from Salisbury to Woking, and drove thence to the Oatlands Park Hotel, near Weybridge : visited Weybridge Church.*

Wednesday, September 8th.

Arrived at 48 Eaton Place.

Visit in the evening from Mr. Hutchings and Sir Edward Baker.

Thursday, Seplember 9th.

Drove with Fanny ; we visited Mr. Boxall.

* His grandmother and great aunt, Mrs. Bunbury and Mrs. Gwyn and their mother Mrs. Horneck are all buried here, and also the Duchess of York, who was much attached to Mrs. Bunbury.—(F. J. B.)

LETTER.

48, Eaton Place,
September 9th, 1869.

My Dear Henry,

1869.

We are stopping here for a day on our way home, because Fanny wanted to get a good deal of shopping done. We have completed our tour very successfully, and arrived so far all safe and well, thank God.

At Salisbury, we received the news of Mr. Pellew's death, and—what shocked me much, as so unexpected—we saw in the papers, the death of Mr. Egerton, of whom and whose family we saw a good deal while we were in town last spring. They had a house almost opposite to us. Few people on so short acquaintance had made a more agreeable impression upon me; and we hoped to cultivate the acquaintance in fact we were expecting to receive them at Barton in October. They were particuler friends of the Kingsleys.

Our tour was very pleasant and we were wonderfully fortunate in the weather; though I do not know that the farmers of the country would rejoice in it so much as we did, for even at the beginning of the time, Sir John Kennaway was in despair about his turnips. It was strange to see the pastures in Devonshire so much burnt up.

I observed that the common oak of North Devon and North-west Somerset is the same as in Wales, the Quercus sessiliflora. In Sir John Kennaway's

grounds I saw the largest liquidambar I ever set 1869.
eyes on, quite a timber tree ; and also a very fine
cork tree.   In the garden of the Bishop's palace at
Wells, a weeping broad-leaved elm, and an Ailantus
both of remarkable size.   Have you a *Paulownia* at
Abergwynant ?   I should think it might do well
with you.

In the Bishop's garden at Wells it seems to flower
regularly in the spring.

I hope we shall be at Barton to-morrow afternoon
and pleasant as our tour has been, I shall be glad to
be at home.

With much love to dear Cissy and your children,

Ever your very affectionate brother,

CHARLES J. F. BUNBURY.

JOURNAL.

Friday, September 10th.

Down to Barton.   Arrived safe and well at our
dear home, and found all right, thank God.

Saturday, 11th September.

We went to look at the new cottage.   Clement
arrived from Scotland.

Tuesday, September 14th.

My Barton rent audit.   Very satisfactory.   After
it I entertained my tenants and Mr. Percy Smith
at luncheon.   Visits from the Arthur Herveys and
Lady Cullum.   Lady Louisa Kerr arrived.

LETTER.

My Dear Mary,

1869.

     I am very glad you have enjoyed such a delightful tour in the Highands. It must indeed have been very interesting, and I hope to hear in due time some of the geological results, for I know Lyell never travels without *results*.

We returned safely and happily from our very pleasant little western tour, last Friday, and we were very fortunate to get home just before the complete change of weather.

I spent nearly two hours of the morning of Monday, the 6th, in the Blackmore Museum, at Salisbury, and was very much interested. Mrs. Fowler brought me into communication with Mr. Stevens, who showed me the whole collection, and gave me an immense amount of curious information—much more I am sorry to say, then I have been able to carry in my head. He is certainly a very clever man, and his learning on that special subject of post tertiary geology and pre-historic antiquities, appears wonderful. In fact, the subject seems to have grown into a science in itself—new as the study is.

Since we came home I have found in the Geological Transactions a description (by Sedgwick and Murchison) of my favourite pebble ridge near Bideford; they consider it as one of the monuments

of a recent rise of the land. We have had a furious 1869.
gale of wind, on Sunday and Monday, which seems
to have been very extensively mischievous ; but here
I am happy to say, we have escaped very well ;
none of my favourite trees have suffered at all
seriously ; but the ground was perfectly covered
with leaves and twigs.

We are both well, and I believe all the better for
our tour ; and I trust you and Lyell will also have
gained in health. My love to him.

Ever your very affectionate brother,
CHARLES J. F. BUNBURY.

JOURNAL.

Friday, September 17th

Mrs. Way* arrived. We walked round the
grounds with her and Lady Louisa.

Tuesday, September, 21st.

Patience, Caroline and Sydenham Hervey came
to stay with us.

Wednesday, September 22nd.

Went to Bury to a gaol committee meeting, and
to the petty sessions.

Heard of the death of poor old George Jones, the
painter. He was an old man, in his 84th year
and had long been in a feeble state yet he sent
three pictures to the Royal Academy Exhibition

* Miss Stanley of Alderley.

1869. this very year, and I believe he went on with his art almost to the last He seems to have died very tranquilly, and without pain. He was a kind-hearted, gentle, courteous, amiable man.

Thursday, September 23rd.

A large barley stack belonging to Mr. Lofts, was burnt down last night—a very disagreeable affair— we went to see Mr. Lofts, and to consult with Mr. Philips.

We dined at Hardwick.

Saturday, September 25th.

Another beautiful day. Fanny and Lady Louisa went to Mildenhall; returned to dinner. The young Herveys went away.

Sunday, September 26th.

We went to morning Church and heard an admirable sermon from Mr. Percy Smith.

I had a long and interesting letter from Lyell on the geology of the Highlands of Scotland, which he has been examining. Of Glen Roy, he says decidedly that "*there can be no doubt* that the glacial theory is the only feasible one," that is the theory that the *roads* were the beaches of fresh-water lakes, dammed up by glaciers. He says, "as " to the parallel roads of Glen Roy, I was agreeably " surprised in one respect. When I visited them "just fifty-one years ago in company with Dr. " Buckland, I remember we found great difficulty in " determining the width and dimensions of some of

" those which when seen from the distance of
" some miles, cut the side of the hill as distinctly as
" telegraph wires often intersect a landscape: but
" when Leonard and I climbed up to the upper-
" most shelf, 1440 feet above the level of the sea, we
" found it perfectly distinct, and for half a mile very
" uniform in width, about ten paces : and so of
" the second shelf, 80 feet below, which on the
" same hill sides sloping at an angle of 31 deg. was
" also 10 paces wide—the roads sloping at an angle
" of about 10 deg. They were admirably defined,
" and when we came to lateral gullies or ravines,
" the roads bent round and entered them for a
" certain depth, as they ought to do if the old lakes
" were sufficiently modern to belong to the present
" glen, when it was almost exactly of the same form
" as it is now, even as to the dimensions of the side
" valleys, with this difference only, that the streams
" which every winter are cutting their channels some-
" what wider and deeper have enlarged the gullies and
" cut away some of the old shelves or beach terraces
" since the time when the barriers of the lakes
" gave way. There can be no doubt that the
" glacial theory is the only feasible one. Yet it
" requires us to assume immense glaciers coming
" from Ben Nevis, and blockages of ice in the
" Caledonian Canal, 1500 feet high, and the lakes
" of that canal, which are deeper than the German
" Ocean, choked up with ice ; and since this glacial
" state of things no derangement in the level of the
" *roads*, though the most remote parts of some
" of them are 25 miles distant from other parts.

1869.   " They who speculate on this subject seem pretty
" well agreed that the *roads* do not belong to the
" first or continental period, when Scotland stood
" higher above the sea than it does now, and was
" wrapped in a winding sheet of the ice : — nor
" to the second period of submergence, when
" Scotland and England were an archipelago, and
" when marine shells were at some points deposited
" in the glacial drift. The shelves are assigned to
" the end of the *third* period after the re-emergence of
" the land nearly to its original height, and when there
" were separate glaciers in the mountain valleys."

In the same letter he tells me that Leonard has
just come back from the Clova Mountains, having
discovered some traces of organic remains in the
metamorphic limestones associated with the
micaceous and gneiss rocks which Murchison
supposes to be Lower Silurian. " His object "
(Lyell continues) "was to ascertain whether the
" two tarns (which you remember) called Loch
" Brandy and Loch Whorral, were rock-basins, or
" merely the heads of small glens dammed up by
" glacier moraines, and which would be entirely
" drained if the moraines of loose matter at the
" outlet were removed. He has come back quite
" satisfied that the latter view is the correct one."

Charles Lyell is now in his 72nd year, but his
mind is delightfully fresh and active, and it is a
pleasure to see that he still retains enough of bodily
vigour and activity to be able to ascend the High-
land mountains. May he long be spared to us.

Monday, September 27th.   1869.

Lady Louisa Kerr and George Napier went
away. We drove to Euston : I called on the Duke
of Grafton, but he was out ; we saw the Phippses,
the Morgans and the Marshalls.

———

Tuesday, September 28th.

I went with Arthur Hervey in our brougham to
Ipswich to a meeting of Governors at the Albert
College—meeting small and business not interesting: a very pleasant talk with Hervey in going and
returning.

———

Wednesday, September 29th.

We drove to Ickworth—saw Lady Arthur and the
dear girls and Lord John : a pleasant talk.

═══

LETTER.

Barton,
September 29th, 69.

My Dear Lyell,

Very many thanks for your letter from
Scotland, which I have read over more than once
with great interest and satisfaction. I am delighted
that your tour has been so agreeable and so
productive of observations, and especially that you
are still able to climb high mountains. What you
tell me about the old Silurian, Cambrian and
Laurentian rocks in the N. W. of Scotland, seems
to confirm Murchison's description of them, which I

H

1869. have just been looking at again in the last edition of Siluria, where he gives also a small sketch of the forms of those remarkable mountains that you mention. It is certainly very curious that the lower beds of the lower Silurian of the Highlands should be in so much less metamorphic a state than the upper, and that the Cambrian sandstones should be horizontal and so similar to Old Red. I do not quite make out from Murchison's description, whether the old Laurentian gneiss is different *mineralogically* from the newer and more extensive Silurian gneiss, nor whether it is stratified as well as laminated.

I was particularly interested by your observations on the *parallel roads* of Glen Roy. I have lately read Nicol's paper on them in the *Quarterly Journal* of the Geological Society, and looked back to McCullock's outlines of the scenery in the old transactions. I was very desirous to know whether, after a personal re-examination of them, you adopted the *lake* theory or the *arm of the sea* theory.

I perceive clearly (I think) from your letter, that the theory of freshwater lakes dammed up by glaciers, is the one which best satisfies you: and that you do not think Nicol's objections valid. I hope we shall have a paper from you on the subject. What I do not understand on any of the theories, and have never seen explained, is *why* this curious phenomenon is confined to one locality? Why it is not repeated in various other Scottish glens? But, I suppose, the answer to this question depends on peculiarities of physical geography, readily understood by those who intimately know the locality and not otherwise.

I was much pleased also to hear of Leonard's observations on the lochs in the Clova district. He is getting a most capital training in geology, and will, I have no doubt, distinguish himself. It must be a great pleasure to you to have the company of such an interesting pupil, and an immense advantage to him.

With much love to dear Mary, believe me ever,

Yours affectionately,

CHARLES J. F. BUNBURY.

JOURNAL.

October 1st.

Yesterday we drove to Drinkstone (near Woolpit), to visit Mr. Harcourt Powell and spend the afternoon in seeing his gardens and his very interesting collection of photographs. He is a very skilful photographer, and his photographs from specimens of pines and firs and other conifers are numerous and very beautiful. He showed us also many excellent photographs of scenery, taken in Scotland and Norway. His collection of living conifers is rich, and includes many fine specimens, especially of Abies nobilis, Thuia Lobii, Cupressus macrocarpa. He has also some good ferns and tropical plants:—Passiflora quadrangularis larger than I ever saw it in Brazil, and bearing fruit in profusion.

1869.   Soon after I had written what is above, the
Arthur Herveys called, to bring us the great and
exciting news that he (Lord Arthur) is to be Bishop
of Bath and Wells.*   He showed me the letter
which Mr. Gladstone had written to him, making
the proposal ; a letter very gracefully and kindly
expressed, and very honourable to the writer.   It is
very agitating intelligence ; I cannot help a great
conflict of feeling.   I rejoice most sincerely in the
prosperity of our dear friends, for whom it is a
delightful piece of good fortune : and I rejoice too
in so admirable an appointment for the sake of the
Church and the public.   I do not believe that a
better choice could possibly have been made.   But
on the other hand, to *us* it will be a terrible loss, an
irreparable loss ; no other man who is or is likely to
be our neighbour can ever supply to me, in any
degree, the place of Arthur Hervey,   There are so
very few *men* who are thoroughly congenial to me,
that the removal of one of those few to a distance is
a most heavy loss.   And indeed, though there are
still two or three other agreeable women in our
neighbourhood, they cannot compensate for the
blank which the departure of Lady Arthur and
her charming girls will leave.   In a public view
too, Arthur Hervey's removal will be an incalculable
loss to this part of the country.

---

* See Journal, August 12th.

Saturday, October 2nd.

Dear Minnie and Sarah arrived—very glad to see them. Mr. Abraham dined with us.

Sunday, October 3rd.

We went (with Minnie and Sarah) to morning Church—a thanksgiving sermon for the harvest, and a collection : we received the Communion.

Monday, October 4th.

Had a very pleasant walk with Minnie, and afterwards with Sarah Napier and Kate Hervey.

LETTER.

Barton,
October 4th, 1869.

My dear Katharine,

I thank you very much for your letter of the 24th, and I was much interested by the information it gave about your family party, and about your visit to Clova, I am delighted to hear of the good progress made in printing your book. At present my thoughts and Fanny's are very full of the appointment of Lord Arthur Hervey to be Bishop of Bath and Wells.

Now to answer your question :—the fine Datura which I am so fond of, is Datura *arborea*,—called in Loudon's Encyclopædia (and often by gardeners), *Brugmansia suaveolens;* but the name Brugmansia properly belongs to a very different genus indeed, next of kin to *Rafflesia.* The Datura arborea is

1869. a special favourite with me, not only for its real
beauty and fragrance, but because its scent always
brings back reminiscences of Rio de Janeiro, where
it grows in every garden.   At Funchal in Madeira,
too, we saw its flowers hanging over all the garden
walls; and it is naturalized in that island.   I believe
it is easily cultivated.   Mr. De Grey has sent me a
specimen of the *Struthiopteris*, lately found in a fir-
wood at Cressingham, between Thetford and Swaff-
ham in Norfolk.   It is perhaps possible that it may
have arisen from spores carried by the wind from a
cultivated plant.   The specimen seems to agree
with the characters of the variety which Willdenow
calls Str. Pennsylvanica, rather than with his Str.
Germanica.

I well remember Clova.   How I should like to
botanize there again, though my walking powers are
by no means now what they were in '44.   I think
there are few things in life more delightful than
a good day's botanizing in a fine country, when one
is in good health and strength.

Fanny is pretty well and very busy arranging for
our dance next Thursday, and for the Bury ball
on Friday.   Dear Minnie and Sarah Napier are
with us.

Pray read Charles Kingsley's article in the
October number of *Macmillan*, on "Woman and
Politics."   I am delighted with it, though he praises
Mill's book* more unreasonably than I should do.

Ever your very affectionate Brother,

CHARLES J. F. BUNBURY.

---

* John Stuart Mill's "Subjection of Women."

JOURNAL.

Had a very pleasant walk with the two dear girls: 1869. Mr. and Mrs. Mills and Ladies Octavia and Whilhelmina Legge arrived.

———

Mr. and Miss Powys, John Hervey, Lord Thurlow and Captain Anstruther arrived.

———

Mr. Powys was obliged to go away.

I walked round the garden and arboretum with Mr. Mills.

Our dance—very pleasant—lovely girls. I danced Sir Roger de Coverley with Lady Cullum.

———

A long talk with Scott, just returned from Abergwynant.

Henry sent me, by Scott, specimens with cones of Taxodium distichum, Cupressus Lawsoniana, and Thuia Lobbii, grown at Abergwynant.

General and Mrs. Ellice dined with us—afterwards Fanny and all the party, except Mr. and Mrs. Mills, and I, went to the Bury ball.

———

Saturday, October 9th.

A most delicious day. John Hervey and three other men went away. William Napier arrived.

The Arthur Herveys dined with us—a pleasant evening. Dear Kate went back with them.

<hr/>

Monday, October 11th.

Most beautiful weather. Fanny and Mrs. Mills went to Mildenhall, returning to dinner.

<hr/>

Tuesday, October 12th.

Dear Lady Napier arrived; also Mrs. Ellice and her daughter Cissy, and George Napier.

We went (a large party) to the Bury Athenæum, to hear Arthur Hervey's lecture on "Wellington and Waterloo:" the last inaugural lecture that he will give at Bury! Interesting in itself, but much more interesting from the circumstances. The concluding part, in which he took his farewell of the institution which has depended so entirely upon him, was very touching.

I feel more and more how great a misfortune the departure of the Arthur Herveys will be to *us*. But I must try to think more of their gain than of our loss.

<hr/>

October 13th.

A very gay and pleasant party in our house since the 6th;—nearly the same as about the same time last year, but without Sarah Hervey (who is detained by company in *their* house), and with

addition of Miss Mary Powys, a very pretty and 1869.
clever and agreeable girl. Dear Kate Hervey with
us part of the time. Our dance on the 7th very
merry and successful. Dear Sarah Hervey arrived.
General and Mrs. Ellice and the Percy Smiths also
came to dinner.

A very merry evening, with extempore dancing.

Thursday, October 14th.

Took a walk with Minnie and Cissy Ellice—very
pleasant.

The dear Arthur Herveys dined with us : a merry
evening.

We received the sad news of the death of that
most amiable, gentle and kind-hearted person, Mrs.
Hutchings. Her poor husband!—I am very *very*
sorry for him.

Friday, October 15th.

The Millses and the Ladies Legge and Minnie
Powys went away.

I had a pleasant walk with Sarah Napier and
Cissy Ellice, and afterwards showed the fern-house
to Mrs. Ellice.

Lady Cullum dined with us.

Saturday, October 16th.

William Napier, Captain Spencer and Clement
went away.

Tuesday, October 19th.

George Napier went away.

1869.   Wednesday, October 20th.

Walked round to see the effects of the gale and
the frost : met Minnie and Lady Napier, and
walked with them.

———————

Friday, October 22nd.

Fanny and Minnie went to Mildenhall to see the
Campbells, and returned to dinner.

———————

Sunday, October 24th.

News of the death of Lord Derby.

Since there has lately been such a clamour about
the appointment of Dr. Temple to a Bishopric, I
have lately read over again his Essay "On the
Education of the World," the first in the volume of
Essays and Reviews. It does not strike me as very
powerful ; it is perhaps a little fanciful, and here
and there certainly rather obscure : but I cannot
conceive what anyone can find in it seriously
objectionable.

———————.

Monday, October 25th.

Sir Henry and Lady Rich and Edward arrived.—
I walked round the arboretum with Sir Henry.

Lady Napier went away.

———————

Tuesday, October 26th.

To the Quarter Sessions at Bury—a great muster
of magistrates—much discussion.

Mr. and Lady Mary and Miss Powys, Lady

Rayleigh, Miss Strutt and John Strutt and Sir E. 1869.
Campbell arrived. The dear Arthur Herveys dined
with us.

* * *

Excessively cold : a furious *snow*-storm in afternoon. A visit from Sir Charles Murray—very
agreeable. The young people very merry with
various games.

* * *

The weather extraordinary. The 24th to the
26th extremely cold, with a bitter and violent north
wind. Yesterday morning still colder, but bright
and fine : but about 2 p.m. came on a furious snow-
storm,—a storm of real *snow*, not mere sleet, which
lasted till near dark. This morning when we got
up, the ground was still perfectly white ; about 9 a.m.
another snow-storm like that of yesterday, came on
and lasted till some time after noon.

* * *

A change of weather ; the day fine (except some
rain in the morning), and rather mild. Mr. and
Mrs. Hardcastle came to luncheon, and the Arthur
Herveys called. Mr. Powys sang in the evening
very agreeably.

* * *

Our pleasant party broke up : all our guests
departed except Minnie and Sarah and Edward. I
wrote to Katharine.

My beloved wife went to London to see her dear friend, Emily Napier on the eve of her departure for Cannes. In the last ten years, we have not once been separated for two days together ; very seldom for one.

Wednesday, November 3rd.

Fanny returned home about eight in the evening, —quite well, thank God.

Thursday, November 4th.

The Frederick Freemans, Sir John Kennaway and his sister and daughter, and Mr. Lott arrived.

Friday, November 5th.

Played one game of billiards with Miss Kennaway. Sarah returned. M. and Mme. and Miss de Bunsen arrived.

The Abrahams and Mr. Sutton dined with us.

Saturday, November 6th.

Showed the arboretum and ferns to Mme. de Bunsen. The Kennaways and Mr. Lott went away. The Miss Richardsons arrived. Also Clement and Charles Strutt. Charming singing in the evening.

Sunday, November 7th.

We went to morning Church, and received the Communion.

Monday, November 8th. 1869.

A sad event has happened. Sir Henry and Lady
Rich were here with us from the 25th to the 30th of
last month, while we had a numerous and very
lively and agreeable party in the house ; Sir Henry,
though he did not look strong, seemed cheerful, and
did not complain of ill health. They left us on the
30th of October, and as they were going away he said
to me that he had spent a very pleasant time here.
As they reached Cambridge, Sir Henry was attacked
with sudden illness, and was with some difficulty
moved to the Bull inn. There he lay, mostly in a
state of insensibility or nearly so, for a week, and in
the evening of Friday, the 5th of November, he
died. Truly, indeed, "In the midst of life we are
in death." It is the more startling because, as far
as I can learn, this death was not caused by any
positive or definite illness, it seemed like a sudden
failure of the vital powers.*

Wednesday, November 10th.

The De Bunsens went away.
The Arthur Herveys and Mr. and Mrs. Hardcastle
dined with us. Matthew Arnold arrived late.

Thursday, November 11th.

Fanny went with Mr. Arnold to Mildenhall, to
the school inspection. Dear Minnie and Sarah
went away ; also Edward.

* We have heard since from persons who knew them well, that both he
and Lady Rich had for a long time been aware that he might die any
minute.

1869.   Mr. Clarke* (Trinity College) arrived—he and
Arnold very pleasant in the evening.

Yesterday's *Times* contained a long letter from
Dr. Livingstone (to Lord Clarendon), but far from
recent, being dated in July '68.   It contains many
curious particulars about the country through which
he has been travelling, but one can hardly trace his
course on any map ; only it seems clear and certain
that, at that time, he had not yet succeeded in
reaching Lake Tanganyika from the south, nor
(consequently), in ascertaining whether there is any
communication between *that* lake and either of the
two which are called Nyanza.   One fact is brought
out very strikingly by the observations of Livingstone, as well as of Speke, Grant and Baker : the
great contrast between the reality of Central Africa,
and the ancient traditional imaginations relating to
it.   Instead of being a desert, like so much of the
northern and southern part of that continent, this
central part is found to be a fine and fertile land, of
great variety of surface, abounding in lakes and
rivers : and in fact, the chief impediments to travelling, arise from an *excess* of water.

————

Friday, November 12th

Mr. Clark and Mr. Arnold went away.

————

Saturday, November 13th.

The Miss Richardsons went away.

The poor little dog Whisky died, to our great
sorrow.

* The Public Orator.

The *congé d'élire*, for the appointment of Lord 1869. Arthur Hervey to the See of Bath and Wells, appears in to-day's *Times*

———

Wednesday, November 17th.

Walked with Fanny, and we decided on the plan for a new vine-house.*

———

Thursday, November 18th.

We went to Ickworth, to stay with our dear Arthur Herveys:—a party at dinner—Sir Edward and Lady Gage, General and Mrs. Ellice, Mr. Grant and others.

Mr. Sam. Smith writes to Fanny, that he has talked with Mr. W. Greg, who has lately been in Ireland: and that Mr. Greg describes the state of that country to be such, that, under colour of Fenianism, any man, whether Fenian or not, can with impunity murder any other man, landlord or not, for any offence, agrarian or not. The overthrow of the Church, has not, he says, conciliated one human being.

———

Friday, November 9th.

At the Rectory at Ickworth. A beautiful day: we had a delightful walk with Lord Arthur and the girls. Fanny went to the top of the big house with Lord Arthur. A most agreeable evening of con-

———

* It was a question whether we should build a temporary or permanent vine-house, and he decided on the latter, as he said that though his successors might not care for flowers, they would be certain to appreciate grapes.— (F.J.B.)

1869. versation.   Arthur Hervey unanimously elected
Bishop of Bath and Wells.

Sunday, November 21st.

We returned yesterday from Ickworth, where we
had been spending two very agreeable days (one
whole day and an evening and morning) with our
dear Arthur Herveys.   The remembrance of our
intercourse with them during these years will always
be among the most prized treasures of memory.
It is a great comfort that my conscience does
not accuse me of having neglected or been un-
thankful for the blessings of their society.

Tuesday, November 23rd.

We dined at Hardwick ; met the Arthur Herveys,
the Abrahams, General and Mrs. Ellice, the Chapmans ; a pleasant party.

Lady Cullum says she derives *testimonial* from
testy, because people are apt to be testy when asked
to subscribe; and no wonder , considering how
common are testimonials now-a-days.

Wednesday, November 24th.

Mr. Isaacson came from Mildenhall, and he and
Scott had luncheon with us, and we discussed several
questions about the Mildenhall Workhouse and
paupers.

We are unhappy about the very alarming illness
of the Archbishop of Canterbury.   His loss would

be a terrible one. The case does not seem to be
entirely hopeless, but I very much fear that the
probabilities are against his recovery. There is
comparatively recent intelligence of Dr. Livingstone.
According to a telegram, a letter from him has been
received at Bombay, dated from Ujiji, May 13th,
1869. Ujiji is on the eastern shore of Lake Tanganyika, so that he must have effected his great object
of reaching that lake from the southward, and will
doubtless have since ascertained whether any waters
flow from it to either of the Nyanzas.

LETTER.

Barton,
Nov. 24th, 1869.

My dear Katharine,

  I have not written to you for nearly a
month, not from forgetfulness, but because I really
had nothing to tell you beyond what Fanny's
journals would inform you of. Nor have I much
more now; though the time is far from being dull
or hanging on hand, it has not lately added much to
my stock of ideas. We spent two very pleasant
days (from Thursday evening to Saturday noon) last
week with the Arthur Herveys; it is delightful to see
them in their own home, such a pattern of a good
and happy family. We met them again at dinner
yesterday, at Hardwick, and I prize every opportunity of seeing something of them before they leave
us, the more as the time is running short, for they
have fixed their departure, I believe, for the 14th

1869. of December.   His successor in the living of
Horringer is, it seems to be, Mr. Burgess, a learned
man, I understand, and of some distinction as a
classical scholar ; Lady Cullum knew him formerly
at Rome.

Lord John Hervey will probably succeed his
uncle as president of the Bury Athenæum, and is, I
should think, a very fit man for it.

I grieve very much for the illness of the Archbishop, and very much fear he will not recover :
though indeed the report this moning (25th) is a
shade better ; and while there is life there is hope.
His death would be a most dreadful loss to the
Church and to the country, an irreparable loss ; I
do not know of anyone (who is at all of a standing to
justify the choice in the eyes of the public) who
could even moderately well supply his place.   Personally too, he would be sadly missed ; I have not,
myself, known him at all intimately, but I have the
impression of his being a very loveable, as well as
estimable man.   Fanny, of course, feels more than
I can, and so will you, on account of the warm
feeling that always subsisted between him and Mr.
Horner.

I rejoice in the news of Dr. Livingstone, and I
now reallly hope to hear of his safe arrival in
England, with conclusive information as to the connexion between the great African lakes, and the
real head waters of the Nile.   How delightful for
Sir Roderick Murchison !

We are going this evening to an amateur concert
at the Athenæum ; a sort of farewell of the Herveys

to Bury, and therefore very interesting to me. I am 1869.
reading Burnet's "History of His Own Time" (of
which I read the first half in the spring) and Adam
Smith's "Theory of Moral Sentiments," and have
also begun to read "*Les femmes Studieuses*," by the
Bishop of Orleans, which Fanny has just read and
admires.

That same Monseigneur Dupanloup is becoming
an important personage, and I hope he means to be
another Bossuet, fighting for the liberties of the Gal-
lican Church against Ultramontanism.

I hope your Geographical Handbook of Ferns is
making good progress ; I was very glad to see it
advertised. I have read nothing very striking in the
way of natural history since Bentham's address. I
go on with the catalogue of my herbarium, and have
been arranging my Cape Geraniaceæ, with the help
of Harvey's Flora, which is remarkably clear and
satisfactory.

With much love to Harry and Rosamond and
your boys.

Believe me ever your affectionate brother,

CHARLES J. F. BUNBURY.

JOURNAL.

Thursday, November 25th.

Wrote memorandums on the Mildenhall Workhouse. An amateur concert at the Athenæum at
Bury ; organised, at the suggestion of Lord Arthur,
to clear off the remaining debt on the institution.

1869. The concert was a great success and very agreeable. The chief performers were — Mrs. Abraham and Sarah and Kate Hervey and a cousin of theirs, Miss Isabel Hervey, Mr. James Bevan, and two of his daughters, Mr. Lyttleton, Mr. Constable, Miss Fearnside.

———.

Friday, November 26th.

Sarah and Kate, with Lady Harriet Hervey and her daughter came to luncheon and spent part of the afternoon with us—very agreeable. A public meeting at the Athenæum, when a piece of plate which had been subscribed for by 200 persons, was presented to Arthur Hervey as a "testimonial" from that institution. They presented at the same time a bracelet to Lady Arthur. Seldom, indeed, can a testimonial have been so well deserved. His speech in acknowledgment was admirably simple and touching; he was much overcome. Mr. Salmon, who spoke for the subscribers, performed his part well.

Grieved by the news of the death of that charming person, Pamela Lady Campbell. She was a kind warm-hearted, excellent woman, and one of the most agreeable I ever knew. I have not indeed seen her very often, since 1836, when I was at Dublin with Edward, and we were almost domesticated in her house at Drumcondra; but the few times that Fanny and I had had opportunities of meeting her of late years, she had shown a very cordial feeling towards us. We are both very fond of her son, Sir Edward.

Saturday, November 28th.

Fanny went to Mildenhall and returned to dinner.

———

Monday, November 29th.

We walked with Scott to see the renovated cottage near the ice-pit ; very satisfactory.

———

Tuesday, November 30th.

Examined Mr. Isaacson's papers relating to the Mildenhall Workhouse, and wrote a long and careful letter to Mr. Longley on it.

———

December 2nd.

We dined with the Gilstraps at Fornham, an agreeable party—the Augustus Herveys, Mr. and Lady Susan Milbank, Lady Cullum, the Praeds, Cyril Wilson, &c.

———

December 3rd.

Sent a donation to the St. George's, Hanover Square Committee for organizing charitable relief. Mr. and Mrs. Percy Smith came to luncheon.

———

Saturday, December 4th.

A visit from Lord and Lady Augustus Hervey.

Weather very severe this week. Snow began to fall on the evening of Tuesday, the 30th of November ; on Wednesday the 1st the ground was well covered with snow, with a hard frost, and in the

1869. night, or early Friday morning, the thermometer in the garden went down to 16 deg, Fahrt.

———— —

December 6th.

I have omitted to notice in its place, that on the 2nd of the month, Charles Kingsley and his daughter Rose sailed for Trinidad—at least all was arranged for their sailing that day, and we have no reason to doubt that they did. I trust they will have a successful and agreeable expedition, and return in safety.

———— ——

Thursday, December 9th.

We went to Mildenhall—saw little Annie and Finetta Campbell—spent the evening in the " red room."

Fanny's favourite collie dog killed a hare near West Stow.

———— ——

Friday, December 10th.

Mildenhall.

Attended the Board of Guardians. Important discussion on question of new workhouse. Mr. Paine's resolution against building new house, carried with only one dissentient voice.

We visited the boys' and girls' schools—very satisfactory.

————

Monday December 13th.

We went to spend the afternoon at Ickworth, a farewell visit to our dear Arthur Herveys. They, that is, the Bishop and Lady Arthur—are to leave 1869. their home on the 15th, and on the 21st he will be consecrated as Bishop, and will cease to be Rector of Ickworth.   The girls will remain at Ickworth till the 20th.   There was a melancholy pleasure in talking with them for the last time in that old home where we have passed so many delightful hours with them.

I could not half express what I felt.   I think they also felt much at parting with us.   To them indeed, of course the *wrench* is much greater; not a separation merely from one family of friends, but from the home of their lives, with all its associations.   I ought to feel more gratitude for having enjoyed for many years the blessings of their society and friendship, than sorrow for the present separation :—

> " 'Tis better to have lov'd and lost,
> Than never to have lov'd at all."

And the loss is, I trust, not total.

Tuesday, December 14th.

A beautiful day.

Walked with Fanny—her collie dog leaped over the arboretum wall.

Sunday, December 19th.

We went to morning Church.   An excellent sermon from Mr. Percy Smith.

A visit from Mr. Percy Smith in the afternoon — a good talk about books.

LETTER.

My Dear Edward,

1869.

I must write you a few lines to wish you a happy Christmas and New Year, though I have hardly anything else to say, for we are leading most quiet and uneventful lives. Only one sad thing has happened here : poor Mr. Sutton, our neighbour, has lost his eldest son, a good and promising young man, who took a very good degree at Cambridge, and of whom they hoped great things. I am very sorry for them. On the 13th we went to Ickworth, to pay our farewell visit to the Herveys ; a melancholy pleasure. Arthur Hervey went to town on Wednesday, Lady Arthur on Friday, and the girls yesterday, and to-day the consecration was to take place. It has been a dismally wet day.

I do not feel at all happy about Henry's health ; he writes in very low spirits, very despondently about himself, and says that Dr. Gull's remedy has done him no good at all, that he finds himself worse than when he came back from London, and that he has left off the arsenic. We must hope that Dr. Gull is right in his opinion, that there is no serious disease, but I do not feel easy about him.

I am deep in Froude's 11th volume, in the horrors of the government of Ireland, which are a match, I think for anything ever charged against the Dutch in South Africa.

I have been delighted with the "Life of Gibson," 1869. one of the very best biographies, I think, that I have read. I am beginning to look over and put into some shape, the papers which our father left respecting Sir John Hawkwood and others, and this leads me to rub up my knowledge of Italian history and geography.

I am very glad to hear that you are making such good progress. We have bought the book of German lithographs from the old pictures in the Boisserée gallery at Munich; a very fine collection, but the very largest and heaviest book now in this house, if not that I ever saw. I have bought another good big book: " Spix and Martius' Travels " in the original, three ponderous quartos with a folio atlas. The English book is a translation only of the first and thinnest volume. Fanny and I are both pretty well. We shall be quite quiet till the 4th January, when we expect the Lyells.

I hope you will spend your Christmas very pleasantly, and I most heartily wish you a happy New Year.

Ever your very affectionate brother,
CHARLES J. F. BUNBURY.

JOURNAL.

Wednesday, December 22nd.

Arthur Hervey was yesterday consecrated in Westminster Abbey, as Bishop of Bath and Wells. May God grant him many years of health and happiness. I feel very sure that his influence,

1869. will be a blessing wherever it reaches.   Dr. Temple
was consecrated at the same time.

Thursday, December 23rd.

Fanny went to Mildenhall and returned to dinner.

Saturday, December 25th.

Read the narrative of the Nativity, both in
Matthew and Luke, in Greek, with Alford's notes.

LETTER.

Barton.

December 28th, '69.

My Dear Katharine,

Leonard arrived yesterday safe and well,
looking well ; he is very cheerful and pleasant, and
we will make him as comfortable as we can, but it is
an unlucky time for him, as we are regularly
snowed up, and geology is as much out of the
question as most other out-door amusements.   I
have not seen such a Christmas since '60, and not
many such since the days of my youth : it is in the
good old orthodox style of snow and frost, and I am
afraid many of the evergreens will suffer.   It is
terribly cold, but I am thankful to say we are both
well, and it is good reading weather.

I have put up for you a small parcel of a few
uncommon English plants, which, perhaps, you may
be glad to add to your herbarium, and I will send

them by Mary. If there are any of them that you
do not want, pray *pass them on* to some other
botanist.

I am very happy to hear that the printing of
your book is finished, and I hope it will not be very
long before I shall see it.

Since the snow came I have not been able to get
to the fern-house, but the last time I was in it, it
was looking well, and the Lygodium scandens was
in especial beauty : (I am not at all sure that it *is*
that species, but that was the name under which we
had it from Veitch). The conservatory is very
gay : Linum trigynum in full blossom, the pretty
scarlet Acanthaceous shrub with the outrageous
name, several Begonias, Cypripedium insigne in
great perfection, a handsome pale purple Cattleya
from Brazil, and several others. We have Christmas roses in flower, but under glass.

For books, I am deep in Froude's 11th volume,
which is a good bulky affair of 670 pages. The
chapter on Ireland is full of horrors, which almost
make one ashamed of one's countrymen. The
English in Ireland in Elizabeth's time, had brought
themselves to look upon the native Irish, not as
enemies to be conquered, but as noxious animals to
be *cleared off*. Froude writes very well, but the
history rather makes me melancholy : it is such a
record of crimes and follies and sufferings and
blunders, with little (except Drake's expedition) of
the redeeming brilliancy and grandeur which we
associate in our minds with the idea of Elizabeth's
reign. It disposes me to think that history, as it is

1869. generally written—even as written by such a man as
Froude, does not give us at all a true idea of a
period. The literature, at least the popular
literature and the memoirs and the private letters
and materials of that sort, would give us, I suspect,
both a truer and a more agreeable picture of a
period than we can get from any records of the
doings of governments.

I was quite delighted with the "Life of Gibson."
I think it is one of the truest and most agreeable
biographies I have read ; it brings the man so
perfectly before us. I am also reading Adam
Smith's "Theory of Moral Sentiments."

What do you think of the Council ? Is not the
Pope a bold man to declare open war against all
modern science, literature, and civilization, and
to expect to bring them all into subservience to
himself ? I think there are hopes that it may lead
to a "disruption" in the Church of Rome. I
heartily wish success to Dupanloup and the Galli-
can party.

(December 31st.)—With all my heart I wish you a
happy New Year, dear Katharine, with health and
every blessing to yourself and all who are dear
to you. I look back with deep thankfulness on the
past year, which has been a *very* happy one to
me ; indeed I cannot feel thankful enough when
I look back on the last five years, which have been
absolutely without a cloud—almost *too* peaceful and
happy for human life. One especial cause for
thankfulness is, that none of our dear and intimate
friends or relations have been taken from us,—

though we have had occasion to sympathise with 1869.
several of our friends in their bereavements. We
are both much pleased with Leonard ; he is a very
interesting young man. With much love to the rest
of your party (whom we hope to see next week),

Believe me,

Your very affectionate brother,

CHARLES J. F. BUNBURY.

P. S. 1. Leonard tells me that in the north-west
of Yorkshire he gathered fifteen different species of
Ferns in one day ; this is an unusual number for
England. The greatest number I ever made a note
of in one day's collecting was 22 ; that was in
Madeira ; but I dare say I might have got a greater
number in one day near Rio de Janeiro, if I had paid
special attention.

P. S. 2.—Fanny desires me to send her love and
best wishes, and to add that she looks forward with
great pleasure to seeing Harry and Rosamond and
Arthur here next week.

JOURNAL.

Wednesday, December 29th.

Snow still lying, but a change of wind.

I have lately begun to examine and put in order
the MSS. which my father left (in a more or less
fragmentary state), relating to military history ;
beginning with the life of Sir John Hawkwood,
about which he had taken much pains, and of which

1869. the MS., though not finished or corrected, admits, I think, of being put into a *presentable* shape. I think that with the pieces intended for his Italian military history, and some extracts from his common-place books connected with the same subject, I may make up a volume to be printed for private circulation, as a supplement or sequel to the Memoir which I printed last year.

————

Thursday, December 30th.

Very rough and blowing weather, excessively cold, though slowly thawing.

The year is drawing near to its close, and again I have to offer up most humble and earnest thanks to Almighty God for the innumerable blessings which I have enjoyed :—above all, for the blessing of having such an incomparable wife always by my side, and for her health and my own. The year which is passing away has been a peaceful and very happy one to me, and, I trust, has not been altogether ill-spent. I feel very thankful that we have been spared from the pang of any losses in our own intimate circle : none of our own loved kindred or dear and especially valued friends have been taken away, though the number of deaths among our acquaintances, or those of whom we knew more or less, has been large, and we have had to sympathize with many of our bereaved friends. Of friends *of the outer circle* and acquaintances, we have lost Mrs. Hutchings, Pamela Lady Campbell, Sir John Boileau, Mr. Egerton, George Jones, Sir Henry Rich and young Sutton.

I have sufficiently recorded in this journal my 1869
feelings concerning Arthur Hervey's appointment to
the See of Bath and Wells, and the consequent
removal of him and his family from our neighbourhood. We can only console ourselves by thinking
more of *their* gain than of *our* loss. I feel very sure
that neither distance nor time will produce anything
like estrangement between us and them.

Charles Kingsley and Rose are now I hope in
Trinidad. We have not yet heard of their arrival.
but I fully hope that they must have been in the
warm latitudes before the last storms set in. This
autumn we were (in common with a very great
number of other people), in great and sorrowful
anxiety about the good Archbishop of Canterbury.
Happily the immediate anxiety has been relieved :
but I very much fear the recurrence of the danger
when he returns to that severe labour which, with
his character and in his position, it will be very
difficult to withhold him.

I very seldom put down in this journal any notes
on public affairs, because they are always worn
threadbare by the newspapers, and I will only say
now, that the political prospect, as concerns Ireland,
and even as concerns England, appears to me far
from pleasant or promising.

1870.

Fanny has had a charming letter from dear Sarah Hervey, describing in a most lively manner their situation and reception at Bath. The house which they have hired for a few months seems to suit them very well.

_____

Sunday, 2nd January.

We went to morning Church and received the Communion. Mr. Percy Smith's sermon very good.

_____

Tuesday, 4th January.

Dear Charles and Mary Lyell arrived.

_____

Friday, 7th January.

Fanny and I went to Mildenhall early. I attended at the police court the examination of two men, Rutterford and Heffer, for murder of a gamekeeper at Eriswell. We had luncheon with the Edward Campbells; returned home to dinner. Harry, Frank, and Rosamond arrived.

_____

Saturday, 8th January.

Visit from Mr. Powys. Charles and Mary speak with delight of their tour in the North-western part of Yorkshire and in Scotland last Autumn, and especially of the beauty of the scenery about Loch

Assynt in Sutherlandshire: the grandeur of the 1870. forms of the mountains and the beauty of the mingled colouring of heath, birch and fern (in their autumnal yellow), and rocks in the foregrounds. With respect to the parallel roads of Glen Roy which he examined very particularly, Lyell says he is convinced that the true theory of these terraces is, that they were the beaches of mountain-lakes which were dammed up by great glaciers descending from the mountains: in the same way as the Märjelen lake in Switzerland, described in the last edition of his "Principles." The only other theory, he thinks, which has even any plausibility, is the marine one—that the roads are ancient sea-margins formed when the country rose progressively out of the sea, which had filled Glen Roy and the glens connected with it. I asked for an explanation of the fact, that these phenomena are so rare and local, not occurring (so far as I understand), anywhere else in Scotland: Lyell replied that *glacier-lakes* are themselves rare and local phenomena. This explanation is not applicable to the *marine* theory, and therefore seems to afford an argument in favour of the other.

Sunday, January 9th.

I walked with Charles Lyell, Rosamond and Arthur to see the family trees.

Charles Lyell told me of a curious discovery of fossil seeds in Scotland. In lately sinking for coal in Ayrshire, they passed first through a consider-

1870. able thickness of boulder clay (what the Scotch call "till,") and then through a bed of a sort of peat or peaty mud, lying between this clay and the coal formation. The curator of the Glasgow museum took away some of this peat, washed it, and found in it a quantity of seeds, which proved to be those of water lily, potamageton and other freshwater plants of the existing flora. He then thought of examining some peaty mud of similar appearance, which adhered to some tusks of Elephas primigenius and antlers of reindeer, that had long been in the museum and were believed to have been found in the same district. On washing this mud also, he discovered in it some seeds of the same kinds. Remains of elephants, Lyell says, are very rare in the quarternary deposits of Scotland. Lyell thinks that the great rarity of elephant remains in Scotland may be owing to the enormous mantle of continental ice which covered that country in the glacial age (as Greenland is covered now), and which may have *ground* away and removed the pre-existing elephant beds.

————

Monday, 10th January.

A beautiful day. Walked with Lyell to the Ice-pit. Dr.,* Mrs. and Miss Phelps arrived.

Charles Lyell tells me that the great question now agitating the United States, is that between the advocates and the opponents of a return to a metallic currency, very much the same question

* Master of Sidney Sussex College.

which was so much disputed in this country for
some years after the great war in which Francis
Horner took an important part; the grounds too on
which it is urged are much the same as on that
occasion, and ignorance of the principles of political
economy is very generally shown now in America,
as it was then in Britain. But the strongest
arguments of the advocates of paper currency, and
a better one than any which the same party with us
had, is that the population and capital of the
country are increasing at such a rate, that the *real*
amount of wealth will in a few years come up to the
fictitious amount represented by the paper. It is
like (as Lyell says), to the case of a suit of clothes
made too large for a boy who is growing fast, he is
sure to grow up to them. Already indeed, through
the operation of this cause, the difference in value
between the paper and the metallic currency has
been considerably diminished. It is in foreign
commerce, Lyell says, that the injurious effect of
the depreciated currency is chiefly felt. The internal
commerce of the States is enormous, and in this the
evil of the depreciation is little felt. Thus the
paper currency to a certain extent acts in effect as a
protection against foreign merchandise : and for this
reason the Massachusetts people, though the most
enlightened in the States, uphold the paper money
system. It is remarkable, Lyell says, that
California, alone of all the States of the American
Union, kept its metallic currency all through the
war and never adopted the paper money system.

The Quarter Sessions at Bury. Alfred Newton and Mr. Lott arrived. The Abrahams, Lady Cullum, Archdeacon and Mr. Chapman dined with us.

———————

Wednesday, 12th January.

The Phelps' went away. The Edward Campbells with Guy and Annie arrived. Lord John Hervey and Patrick Blake dined with us.

Alfred Newton (the Zoologist) says that Landseer's great picture of Swans attacked by Eagles is "a marvel of untruthfulness." As to his other picture in the same exhibition—the Ptarmigans, Newton says that the changes of plumage in those birds do take place slowly and with some irregularity, so that it is not absolutely impossible that different birds in the same flock might be in different states of plumage at the same time.

———————

Thursday, 13th January.

Walked with Lady Campbell, Mary, Fanny, Charles Lyell and Alfred Newton about the premises.

Fanny has had a delightful letter from Kate Hervey, describing in a most lively manner their visit to Wells, and the enthronement of the Bishop He seems to have been received there with the enthusiasm he deserved.

Alfred Newton said to-day, that he has observed birds not to be so constant as is commonly supposed in the choice of places for their nests, and to be often influenced by circumstances. For instance, in Lapland he observed numerous nests of the Peregrine Falcon on low sallow bushes, hardly a yard high, in the midst of extensive marshes, where the birds find abundance of food : and in so unpeopled a country they are exposed to very little danger from man. In Britain, and in Europe generally the Peregrine Falcon nests always amidst lofty and precipitous rocks, especially sea-cliffs. The Golden Eagle again, in Britain, makes its nest always on lofty crags:—in Turkey, often on low trees, so that a friend of his in that country actually took an egg out of an Eagle's nest without dismounting from his horse.

Newton does not think that the Golden Eagle is in present danger of extinction in Scotland. The great landowners, especially owners of deer forests, takes some pains to preserve it, particularly to preserve its nests, and he thinks that the number of the species, in Scotland, have rather increased than diminished of late years. The Sea Eagle, he thinks, is more exposed to persecution.

Friday, 14th January.

Dear Charles and Mary went away ; also Alfred Newton.

15th January.

I had a pleasant walk with Edward Campbell. He thinks Montague MacMurdo the ablest military man in this country, and the one, who, if we were

1870. engaged in a serious war, would soon rise to the
head. This is exactly my opinion also.

Campbell showed me an interesting series of
photographs of buildings and scenery in India,
particularly at Delhi and Agra.

Among them were two views of a famous wood of
Deodars near Simlah, very striking : the trees so
different in form and habit of growth from Deodars
as we know them in cultivation, that one could
not have guessed them to be the same kind ; doubtless from having grown closely, and being thus
drawn up, they have immensely tall, straight, mast
like trunks, and branches comparatively few and
short ; in fact, one would take them for Scotch firs
rather than Deodars.

Sir Edward and Lady Campbell are both charming, and I believe most estimable, indeed admirable.

January 17th, Monday.

We have had distressing news of the dangerous
illness of Frederick Freeman : I fear his state is
very alarming.

January 18th, Tuesday

The Campbells left us. The telegraph this
afternoon brought us the sad news of poor Frederick
Freeman's death. I am much grieved at it. My
acquaintance with him began in January, 1839, just
31 years ago, at the Cape, whither he came out with
his mother. To that poor mother the blow of his
loss will be dreadful. His excellent wife was
devoted to him, but I think she was better prepared

for this fatal result, and she is certainly of a very 1870.
calm and firm nature ; she will bear the blow as well
as anyone could ; and she is, I do not doubt, admirably fitted to have the charge of her boys.

Mr. and Lady Susan Milbank, and Lady Emma
Osborne, John Hervey, Frank Lyell, and Herbert
arrived, and later, dear Catty Napier, and the
Bruce girls, Rachel and Jessie.*

Pleasant talk with Catty. The Thornhills arrived.
They and most of our party (but not Fanny or
Catty or I) went to the Bury Ball.

The Milbanks, Thornhills and John Hervey went
away.

Fanny and Catty went to spend the day at
Mildenhall.

Mr. and Mrs. Longley arrived.

---

* Rachel became the wife of Augustus Vernon Harcourt, and Jessie, the
wife of the Rev. Wynne-Jones, Rector of Cardiff.—(F.J.B.)

Wednesday, 26th January.

Lord John Hervey, Henry Waddington, Maitland, and Henry Wilson dined with us.

Friday, 28th January.

Fanny went with Mr. Longley to Mildenhall.

LETTER.

Barton.
January 28th, 1870.

My Dear Edward,

I have no doubt you have been very much shocked and grieved as we have, by the death of poor Frederick Freeman. We had heard so little of his illness, that the news of his danger came upon us with a sudden shock; for though he looked very ill, when he was here in November, we never dreamed of death being near. He was a man very much to be regretted; exceedingly amiable and warm-hearted; and a most true and constant friend; very sweet tempered and pleasant. I feel very much for his poor wife and mother; but I am very glad to hear that Lady Napier seems to bear the shock better, and to be less overpowered by it; than could have been expected.

We have had a good deal of company this month, and some very pleasant; the Lyells and Alfred Newton, and Dr. and Mrs. Phelps in the earlier part of the month; then the Campbells; and lately dear Catty Napier; who is a special favourite of mine;

indeed so are the Campbells. Herbert Bunbury 1870. and Frank Lyell have also been with us for some time; Clement for a day or two and we expect him back on Monday. To-morrow, I suppose, we shall know where he stands in the list of honours. He does not expect to be high. I was very glad of your information about books. I quite agree with you concerning Froude, whose 11th volume I have lately finished, but I hope that the 12th will be more interesting. I never read any book that gave me so bad an impression of Elizabeth.

I have got Mr. Lear's book on Corsica, but have not yet read any of it; only looked at the prints; the scenery is very wild and magnificent. I forget whether I told you that Mr. Hawker (a clergyman in Hampshire, a zealous naturalist and a member of the Alpine Club) sent me his paper on Corsica, reprinted from the *Alpine Journal* containing several notices of the Botany; and he also sent me seeds of the true Pinus Laricio, which he describes as a magnificent tree, and so it appears in Lear's views.

(January 29th). Clement is 10th Junior Opt., not a splendid position, but it might have been worse.

Ever your affectionate brother,

CHARLES J. F. BUNBURY.

JOURNAL.

Sunday, January 30th.

I received from Katharine her new book on Ferns.

I read Kingsley on Prayer and Science, and on Humility.

1870.   Monday, January 31st.

[During this month he was much engaged with his
Father's MSS. of the Life of Sir John Hawkwood,
which he at that time intended to print.—F. J. B.]

LETTER.

Barton, Bury St. Edmund's,
February 4th, 1870.

My dear Katharine,

Many thanks for your kind letter and good
wishes on my birthday.   I am very thankful to be in
the enjoyment of good health and of so much
happiness at the end of my 61st year, and in the
enjoyment too of the kindness of so many loved and
valued friends.

I gladly accept your book as a birthday present,
and value it very much I assure you.

Fanny has had a charming letter from Rose
Kingsley from Trinidad; they had a favourable
voyage, and are very happy, revelling in ferns and
butterflies, and all the glories of tropical nature.
She says the climbing ferns, in particular, remind
her of the fern house at Barton.

I read Wallace's "Amazons," and "Rio Negro"
when it first came out; it is very good, but it did
not fascinate me nearly so much as Bates's book;
I was struck by its extremely prosaic and matter-of-
fact tone—something of a want of enthusiasm.

Ever your very affectionate Brother,
CHARLES J. F. BUNBURY.

JOURNAL.

My 61st birthday.   Thanks be to God.   1870.

————

The meeting of the Archæological Institute at Bury.   Lord John Hervey elected President— various alterations made in rules.   Walked about the arboretum with Fanny.

I received on the 30th of last month from Katharine Lyell, one of the first copies of her book, "A Geographical Handbook of Ferns."   It is very well done, and very honourable to her industry and accuracy, and will be of great use both to students of Ferns in particular, and of botanical geography.

══ ══

LETTER.

Barton,
February 7th, 70.

My Dearest Cissy,

I owe very many thanks both to you and dear Henry for your kind loving letters and your good wishes on my birthday.   I must thank you heartily, for a pretty and very appropriate book which you have so kindly sent me.   "The Comforts of Old Age" is a subject on which much may be said, and I fully trust and believe that it has many comforts. I am sure, I feel, as I approach it, that I have

1870. numberless causes of thankfulness and happiness, in fact I feel with Thackeray, that it is a very comfortable thing to be an *old fogy*—at any rate when one is blessed with so many to love as I have, and with good health besides.

I wish with all my heart that Henry had the enjoyment of good or tolerable health.

I am very glad that Henry thinks of training Willie for Woolwich and the Engineers: it is, I should think, the very line for a clever spirited lad who can study and who at the same time is of an active and enterprising turn: it affords opportunities both for the intellectual powers and for activity and energy. Our neighbour Captain Horton said to me the other day, that he means to educate his son with a view to Woolwich and the Engineers, and then, whether he ultimately goes into that branch of the profession or not, he will at any rate have acquired a great deal of knowledge which will be very useful to him in any position in life. I think he is very right. I am much interested also in what you tell me about George. As he has a real inclination to be a clergyman, and thinks seriously about it, I think it very right to encourage him in it, and to keep that object in view. It is a noble work when taken up in a right spirit, though it is not one which in these days offer a very inviting prospect in a worldly point of view, including the possible prospect of Disestablishment which looms in the distant (I hope very distant) future. I seriously hope that before George is old enough for Orders, clergymen of the Church of England will be relieved from the burden and snare of subscription to Thirty-nine Articles, 1870. which is now making so many unhappy who subscribed before they knew their own minds, and now find themselves helplessly fettered and entangled by what they cannot believe and cannot escape from. I do believe that this is a very serious source of evil, and that it restrains many able and thoughtful men from entering the Church.

I hope we shall see you *all* here in the summer, before dear Henry goes to sea, and I shall be delighted to pay for the journey of the whole party, as well as for R. N.'s* outfit.

I was very sorry for poor Toby's† accident, but am very glad that he has so well recovered.

With my hearty and earnest love to Henry and your children,

Believe me ever,

Your affectionate brother,

CHARLES J. F. BUNBURY.

JOURNAL.

Monday, 14th February.

Terrible weather. My Barton rent audit—very comfortable and satisfactory.—Afterwards I entertained all my tenants, together with Mr. Percy Smith at luncheon.

Kate MacMurdo arrived.‡

---

* His eldest nephew, Harry, just entering the Navy.

† A dog.

‡ She brought her little brother Arthur, not 9 years old, to go to school at Bury.—(F. J. B.)

Tuesday, 15th February.

The Barnardistons, John Hervey and Captain
Boileau came.   Lord and Lady Augustus Hervey
dined with us.

–––––– ––

Wednesday, 16th February.

A fall of snow in the morning and afterwards a
thaw.   We had at dinner Mr. and Mrs. Abraham
and the Bishop of Wellington, New Zealand, the
Hortons, Percy Smiths, Mr. Wilson and Mr.
Grant.*   The Bishop (Dr. Abraham) is a very
agreeable, cheerful and interesting man.   He
brought with him and showed to us a number of
very beautiful and interesting coloured drawings
(done by his wife) of New Zealand scenery.   A
very fine and picturesque country it seems to be,
and several of them are admirable views of forest
scenery, giving a clear idea of the luxuriant quasi-
tropical vegetation, the great trees hung with huge
woody climbers and draped with epiphytes, almost
as in Brazil, and the fine tree ferns, etc.

The Bishop spoke particularly of one gigantic,
woody epiphyte—a myrtle with crimson flowers
he called it (query a Metrosideros ?) — as one of
the most remarkable plants in the New Zealand
forests.   I did not quite understand from his
description, whether it is strictly an epiphyte, or
originally a climber ; but it seems to embrace and
gradually overwhelm and stifle the tree on which it
has planted itself, and finally to kill the tree and

* Rector of Hitchin, succeeding Professor Henslow.   He was nephew of Lord
    Glenelg, who spent the latter part of his life in his rectory.

flourish on its ruins, exactly as various species of 1870. ferns do in both the East and West Indies.

I have read with some care, Gladstone's speech on introducing the Irish Land Bill, and the analysis of it given in the *Pall Mall Gazette* and *The Times.* The measure is (as it ought to be) specially and distinctly framed with a view to the special circumstances and requirements of Ireland, and therefore I feel myself scarcely able to understand or judge of many of its provisions ; to do so would require an intimate acquaintance with Irish affairs, which I do not possess. As far as I can understand it, I do not perceive anything objectionable in the measure itself. But there is something which I do not quite like in the tone of some parts of Gladstone's introduction ; he seemed rather disposed to speak too lightly of the rights of property.

========

LETTER.

Barton,
February 18th, 1870.

My dear Joanna,

You will have heard from Fanny of the agreeable accounts we have had from the Arthur Herveys since they went away, how happy they are, and how much the people of the diocese are delighted (as well they may be), with their new Bishop and his family. We miss them even more than we expected, but we hope to see a good deal of them in London in the spring. Fanny had a charming

1870. letter from Rose Kingsley from Trinidad, where she and her father were staying with their friends the Governor and his wife.* They had an excellent passage out, were very well and very happy, and as much enchanted with the beauty of the country and the wouders of nature as I had expected they would be. I have heard since from Mrs. Kingsley (who is in England), that Charles Kingsley had gone off with the Governor of Trinidad (but without his daughter), on an expedition up the Orinoco, which will be very interesting to him.

I have been lately dabbling a little in Florentine history ; for I have begun to look over, and put into shape the papers on military history which my father left in MS. ; and the first I have taken up, and on which I am engaged, is the " Life of Sir John Hawkwood." In revising this and adding a few illustrative notes, I have been led to rub up my knowledge of that history. Everything that relates to Florence in the middle ages is interesting, and many of the names of places which I meet with, recal our delightful evening drives in the summer of '66   Kate MacMurdo is with us just now, but is to leave us next week, her father and mother are in London, getting ready in a great hurry, to go out to India for five years. MacMurdo has accepted the command of the division quartered in the Thelum district in North Western India ; his headquarters will be at Rawal Pindee, near where the Indus issues from the mountains. They take Kate and two of the other girls with them, and leave all the rest of

* The Arthur Gordons.

the children in England. My nephew Cecil has 1870. been with us some days, being on leave of absence from Gibraltar : he is a very fine young man indeed. I have been very uneasy this winter about Henry's health, and am by no means entirely comfortable about it yet, though the accounts lately have been better.

Pray give my very sincere and hearty thanks to dear Susan for her most kind and affectionate letter on my birthday.

We shall be very glad to supply you with bulbs of snowdrops, both the common and the large Crimean kind, whenever there is an opportunity. It is curious that the common snowdrop should be unknown in the gardens of Italy, as it grows wild in that country. The only place where I ever saw it really wild was on Monte Cavo (Neons Latialis) near Albano.

I have nearly finished the 12th and last volume of Froude's History of England. He has quite satisfied me that Elizabeth was neither great nor good, and that Mary Stuart had every fine quality which is compatible with an absolute want of moral principle.

With much love to dear Susan,

Believe me ever your very affectionate brother,

CHARLES J. F. BUNBURY.

(February 20th).

JOURNAL.

1870.   Mr. Forster's Education Bill appears to me, as far as I am yet able to judge, a wise and statesmanlike measure, and one which is likely to do a great deal of good.   I approve particularly of the principle upon which he has professedly framed it, of *utilizing* all that has hitherto been done—seeking to complete and supplement and fill up the gaps in the existing work, instead of sweeping all away, and beginning from new foundations in the revolutionary fashion.

Fanny and Kate MacMurdo went to Mildenhall to see the Campbells.

Dear Kate MacMurdo went away—very sorry to part with her.   A visit from Lady Cullum.

A very fine day.   A fire last night at Mr. Lofts" premises at the Charity Farm.   Walked with Fanny—we went to see poor Mr. Lofts.

We went to morning Church—excellent Sermon from Mr. Percy Smith.   Walked with Fanny to look at the memorial trees.

LETTER.

Barton.
March 4th, 1870.

My Dear Mary,

Winter has come back in all his fury; the 1870. fifth act of winter. It has been *snowing* furiously all day, with a violent northerly gale. I wonder whether we shall ever see the end of this winter. I find by reference to my register that the flowering of all the early plants is fully a month later, or more, this year than last. For instance, the common snowdrop flowered in 1869, on January 9th — in 1870, February 20th; the Cloth-of-Gold crocus, on January 16th last year, on February 22nd this year; the common yellow crocus on January 30th in '69, on February 26th in '70, and so on. The Almond tree last year was in full blossom before the 20th of February; this year it is hardly yet even showing a bud.

I have very special cause to be thankful that I have got so far through this very trying winter without even an ordinary cold.

I have just been reading, in *Good Words*, Kingsley's first letter from the tropics, giving his first impression of the West Indies—very interesting.

Did you read the article in the *Edinburgh*, on The Last Two volumes of Froude? and do you happen to know who wrote it? I think it very good. I agree with you in liking Mr. Lecky's first work better than his second, and I hope some of these

1870. days to read it over again, but at present I am
engaged with Dugald Stewart's "Active and Moral
Powers." I am reading to Fanny in the evenings,
Morris's "Earthly Paradise," which I think very
sweet poetry indeed. I use that epithet purposely;
very sweet and tender and graceful, especially in the
stories of Cupid and Psyche and of Alcestis. I
have also, at idle moments, read one volume of
"Red as a Rose," amusing enough, I think.

I see in Sir Roderick's geographical address, that
he is very sanguine about Dr. Livingstone.

I like Mr. Forster's Education Bill on the whole, and
I like it particularly because it is not revolutionary;
it does not begin with making a *clean sweep*, but seeks
to turn to account all that has hitherto been done,
and to extend and improve the existing work. What
I do not like (or rather what I doubt about) is the
leaving on local authorities, the responsibility of
choosing between compulsion and non-compulsion.
If there is to be compulsion, it had better, I should
think, be general and uniform.

I have just bought Hedwig's great Work on
Mosses, the especial *classic* on that subject.

With much love to Lyell and to Katharine,

Believe me ever,

Your very affectionate brother,

CHARLES J. F. BUNBURY.

March 6th.

P.S.—I have not spoken warmly enough of the
" Earthly Paradise," both Fanny and I are quite
delighted with the " Love of Alcestis."

JOURNAL.

March 13th, Sunday.

Read Luke, chapter 24th, in Greek, with Alford's 1870.
notes. Read an excellent sermon by Dr. Temple,
on " Temper."

LETTER.

Barton,
March 14th, 1870.

My Dear Katharine,

We have lately had the pleasure of hearing
that the Kingsleys have returned safe from the
West Indies, and they are all now assembled at
Eversley, including Maurice who has also arrived
safe from Uruguay. I am longing to hear more
particulars, especially about Charles Kingsley's
trip up the Orinoco.

You said in one of your letters that it was difficult
to complete one's collection even of British plants.
Mine is far from complete. I find by a note I made
in Hooker's " British Flora," that at the end of
1844, I had, speaking roughly, about two-thirds of
the species enumerated in that work. I have not
added much to the number since then, except a fine
set of Carices given me by Dr. Balfour, at Edinburgh, in '50. I have not made a complete
calculation of the same kind with reference to
Bentham's " Handbook," but I have done so
with the largest families. I find that of 99

1870. Grasses which he admits (I think he has reduced them rather too much)—I have British specimens of 79 ; in round numbers four-fifths. Of Compositæ I have only 70 out of his 114 ; (I confess they have never been a favourite family with me). Of Legum-inosæ 51, of Bentham's 70. Of Cyperaceæ 50, of his 73. I mean those only of which I have native British specimens. But I have not travelled enough in our own country ; this was one of the many mistakes of my youth.

We have had a delightful quiet and studious time by ourselves, and have read a good deal, though I never find that I get through so much in the day as I wish ; but this week we are going to have two large parties, and next week we are going to stay at Stutton, that I may attend the assizes at Ipswich.

My love to your husband and children.

Believe me ever,

Your very affectionate brother,

CHARLES J. F. BUNBURY.

JOURNAL.

Tuesday, 15th March.

Went to the Quarter Sessions at Bury—many magistrates of my acquaintance—much discussion of county business.

Wednesday, March 16th.

Received a very interesting letter from Charles

Kingsley concerning his late visit to the West Indies. He says, *inter alia:*—"I have seen more "than my wildest dreams could have anticipated. "But I have done much less."——"Of the wonder "of the whole place I will not attempt to write.— "I need not to you, who know the forests of Brazil. "I have to look at times at my specimens to assure "myself it is not all a dream. But in the wonderful "improvement in my health, and in the renewed "youth of my mind, I feel very fair proof of "that—

> "I, too, on honey-dew have fed,
> And ate the fruits of Paradise."

Speaking of his son Maurice, he says :—

"I have oftened longed to get him a post on some "exploring expedition, if such things ever exist "again, under a cheese-paring government, which "fools John Bull to the top of his bent in his "strange fancy that he is over-taxed and must "needs save odd farthings."

The Waddingtons, the Praeds, Lady Hoste, Mr. Abraham, and Colonel Tyrell arrived,—Lady Cullum, the Wilsons, and young Murray dined with us.

LETTER.

Barton,
March 16th, 1870.

My Dear Leonora,

I hope you have none of you suffered from this very long and uncomfortable winter, which has

1870. given us so many repeated doses of snow and
frost, beginning almost before the end of October,
and ending, certainly not yet. We have great
reason to be thankful that we have got through it
hitherto without any illness.

We feel very much the loss of our friends the
Arthur Herveys, and miss them more even than we
expected : but it is a comfort to know that they are
well and happy, and that the Somersetshire people
are delighted, as they ought to be, with their new
Bishop and his family. Our friends the Kingsleys,
that is, Charles Kingsley and his daughter Rose,
are just returned home from Trinidad.

Our Ferns are doing pretty well, and the conservatory is in great beauty at present, gay with
Poinsettias, Eucharis, Camellias, red and white,
many varieties of Epacris, Calla Æthiopica, scarlet
Salvia, Hyacinths, Narcissi and many others.

We enjoyed a very quiet time during the last
week of February and the first thirteen days of this
month, seeing hardly anyone : it was very pleasant
and comfortable and very good for reading—though
I always find that I cannot get through half so
much in the day as I hope to do. I have read the
last two volumes of "Froude," the 11th and 12th
which he has now decided to conclude his history.
The defeat of the Spanish Armada is magnificently
related, but on the whole I do not think that
these two volumes are quite so good as some of the
others : the 11th is certainly tedious. Froude is
still determined to admire Elizabeth : yet his own
narrative gives one the most unfavourable impression

possible of her; she was mean and shabby to a 1870.
degree almost incredible, and in fact seems to have
had nearly every fault that man or woman can
have.

On the other hand, Froude shows an incon-
ceivable degree of bitterness, amounting to absolute
hatred against Mary Stuart, yet it is clear from his
own account, that, though totally without moral
principle, she had many fine and even noble
qualities. I have also read Adam Smith's " Theory
of Moral Sentiments," and am now reading Dugald
Stewart's "Active and Moral Powers," which is
beautifully written.

Give my love to Annie and Dora, who I suppose
are grown quite out of my memory. I should be
very glad to see them again, and I hope you will all
come and spend some time with us this next
summer or autumn.

With affectionate remembrances to your husband,
believe me,

Your affectionate brother,
CHARLES J. F. BUNBURY.

Are you not pleased with Katharine's book?* I
think it excellently well-done.

---

* Geographical catalogue of Ferns.

JOURNAL.

1870.   A beautiful day. Fanny went to Mildenhall to see Mrs. Marr* who is ill. I walked about the park.

––––––

A beautiful day.   We went to morning Church, and afterwards walked about the grounds with little Arthur MacMurdo.

I read a good Sermon by Dr. Temple, on Seasons of Penitence.

––––––

Fanny went again to Mildenhall to see Mrs. Marr.

––––––

Examined the flower of Rogeria cordata, a very pretty cinchonaceous plant, which has just come into flower with us, and wrote a note on it.

––––––

Sir Edward Campbell, his Sister Mrs. Harvey, and his daughter Annie arrived.

––––––

A beautiful spring day.   Out all the morning with Fanny and the Campbells about the park and grounds.

* A housekeeper and kind friend.

Went to the Thingoe Union Board—debate about an alteration of the outer walls. Sir Edward Campbell and Mrs. Harvey and sweet Annie went away. Mrs. Harvey (Fanny Campbell, Sir Edward's half-sister), told me the other day, that where she lives, near Bodmin in Cornwall, the Woodwardia radicans thrives perfectly in the open air, without any protection in winter, and its fronds grow to their full size, between 4 or 5 feet long.

I had not seen Mrs. Harvey since 1836, when she was a girl in her teens (18 I believe), living with her father and stepmother at Drumcondra, near Dublin. She is a clever, well-informed pleasant woman.

Saturday, 2nd April.

Wrote to Sir Edward Campbell to thank him for a leopard skin.*

LETTER.

Barton,
April 2nd, 1870

My Dear Mary,

I have now read through more than half of Lord Stanhope's "Reign of Queen Anne," and can safely recommend it to you, as (to my mind) a very agreeable and satisfactory history, written in a clear, easy, pleasant style, and in a very fair spirit. It is

* Now in the dining room at Barton.

1870. a very different sort of history from Froude's,
never splendid or exciting, never rhetorical, and
never overloaded or tedious,—different also in this,
that he is remarkably good-natured and indulgent,
(sometimes perhaps almost to excess) in his judg-
ments. The military operations, which necessarily
make up much of the history of that reign, are
explained in a satisfactory manner: the battle of
Blenheim indeed, I think, is not quite so well
related as in Creasy's Decisive Battles, and both
that and Ramillies, I understood the better for
having long ago studied their tactical principles in
MacDougall's "Theory of War." But of the operations in Spain, I had not before read so satisfactory
an account. The proceedings connected with the
Union with Scotland are told in a very clear and
lively manner. Truly your Scottish countrymen
were in those days a turbulent race, and there was a
narrow escape from a civil war between the two
kingdoms. The only fault I find with Lord Stanhope is, that his book is in too thick and cumbrous
a volume; I should have liked it better if it had
been made rather longer and divided into two
volumes.

With much love to Charles Lyell and to Katharine,
Believe me ever your affectionate brother,
CHARLES J. F. BUNBURY.

JOURNAL.

We went to afternoon Church.   Read an excellent
sermon by Mr. Rickards, on the Spring.

———

The new vine-house completed.

Edward writes to me :—"I hear that the people
" interested in the Education Bill have good hopes
" of getting it through, notwithstanding the oppo-
" sition of the Dissenters, as there is a very strong
" and general feeling in its favour.   But the op-
" position to the Irish Land Bill is gaining strength,
" and it is feared that the Tories will succeed in
" protracting the struggle against it so long as to
" interfere materially with the prospect of carrying
" the other measure through this session.   One
" cannot wonder at people who look at the Bill from
" the ordinary landlord's point of view being opposed
" to it, but I am told that most of the Irish land-
" lords are contented with it."

———

Went with Fanny in the pony carriage to the
Shrub, and walked through it.   Wood anemones
just coming on.   Studied Mosses.

———

Went out with Fanny in the open carriage, and
visited Lady Hoste, and took her with us to Hardwick.

1870.   Monday, 11th April.

The 26th anniversary of our happy betrothal.

————

Tuesday, 12th April.

Fanny went with Clement and Arthur MacMurdo to Mildenhall, and brought back Annie and Gerald Campbell.

====

LETTER.

Barton.
April 13th, 1870.

My Dear Edward,

Many thanks for your very pleasant letter of the 4th ; I hope you will have a pleasant visit to Devonshire.

As to the Irish Land Bill, I must confess that I should have great difficulty in supporting the Ministers on it. The more it is discussed, the less I like it. The principle of giving compensation for merely surrendering and quitting a farm on the expiration of a lease—which is in effect the principle that mere occupancy gives certain rights—appears to me very mischievous. If it is justifiable at all, it can be only on the ground, that the social state of Ireland is so anomalous and exceptional, that principles must be admitted there which would be inadmissible in any civilized state. I was much interested by your information about the Roman sarcophagus at Westminster,

Having finished Lord Stanhope's " Queen Anne,"

which I liked very much, I am now reading the 1870.
early part of his former work, from the Peace of
Utrecht onwards. I have read it before, but (at
least this early part) many years ago.

We shall not go up to London before May.

Ever your very affectionate brother,
CHARLES J. F. BUNBURY.

JOURNAL.

Thursday, 14th April.

Excursion to Ely with the Campbells. We,
Clement and Annie, went by railway: and the
rest of the party (4) joined us at Kennet. We had
luncheon at the Lamb ; saw the Cathedral; called
on Mrs. Merivale. Home late.

April 17th, Easter Sunday.

A beautiful day. We went to morning Church
and received the Communion.

April 19th, Tuesday.

Lady Hoste and her children and Lady Augusta
Cadogan came to luncheon and spent all the
afternoon with us. We showed them the arboretum, ferns, &c.

LETTER.

Barton,
April 21st, 1870.

1870.   My dear Lyell,

   I thank you much for your letter from
Ilfracombe. I am very glad that you are enjoying
such a delightful tour in North Devonshire; it is
indeed a beautiful country, and most enjoyable in
fine weather such as this; I have seen it three or
four times, but never in spring, and therefore never
saw much of the sea birds which you mention. I
have no doubt the spring flowers in the coombes
must be very beautiful, but on the other hand the
leafless state of the oak woods must be a disadvantage (except with a view to geological sections), and
I suppose the Ferns are hardly yet appearing.

  Your object I presume has been to judge for
yourself of the sections referred to by both
parties in the great Devonian controversy.

  The weather here has been as beautiful as
possible for more than a week past, and we enjoy it
the more after such a long winter. The warm
sunshine has brought vegetation forward so rapidly,
one seems almost to *see* the leaves grow from hour to
hour, especially those of the horse-chesnut and the
hawthorn. The former, I think, will be in blossom
in a few days. I have not yet heard the Cuckoo,
nor has Fanny, though I understand that it has
been heard; I am so deaf that I should not be
likely to be first to hear it. Neither have I yet seen

a swallow, but I have seen peacock, tortoise-shell 1870. and brimstone butterflies.

Give my hearty love and thanks to dear Mary. I am very glad indeed to hear that you are so much better for your tour.

Believe me ever your affectionate friend,
CHARLES J. F. BUNBURY.

P.S.—I have heard the *cuckoo* this afternoon.

JOURNAL.

Monday, 25th April.

The shocking news arrived of the murder of three English gentlemen (Mr. Vyner, Mr. Herbert and Mr. Lloyd) and an Italian Secretary of Legation, by Greek robbers, in the mountains near Athens.

Tuesday April 26th.

The Campbells arrived with several children.

Wednesday, 27th April.

Lady Cullum with Miss Forbes and the Baroness, her niece, Lady Hoste, Mr. Wilson, Mr. Percy Smith, Mr. Murray, and a large number of children of various ages came to play with the young Campbells. A large luncheon and much merry-making.

Sir Edward Campbell went away.

May 3rd, Tuesday.

We went with Annie Campbell to the Athenæum, to Lord John Hervey's lecture on Crabbe—very good.

May 4th, Wednesday.

Fanny took sweet Annie and the rest of the Campbell children back to Mildenhall.

Tuesday, 10th May.

Arrangements preparatory to leaving home.

Wednesday, 11th May.

We went to our dear kind old friends, the Mills at Stutton ; we found them both, I am sorry to say, far from strong, though free from positive present illness. John Hervey (the Rev.) and Mr. and Mrs Anstruther staying in the house.

Saturday, 14th May.

We left our dear friends at Stutton after luncheon.

The weather being very stormy, and having caught cold I was unable to go out while I was there, so could only enjoy the beauty of the place through the windows.

We went up to London.

Sunday, 15th May.

Had visits from Kate Hervey, Minnie and Sarah, Charles and Mary and Edward.

Monday, 16th May.

Frank Lyell and Annie Campbell came in at luncheon. Went to the Athenæum ; saw there Gurney Hoare and Edward Romilly.

Visit from William Napier. Went to the Athenæum — met Edward, Mr. Clark of Dowlais, the Solicitor-General,* John Moore, Douglas Galton —also Mr. Thornhill and Patrick Blake.

Mr. Clark of Dowlais was with us at Barton last year at the time of the Archæological Meeting ; a pleasant, hearty, cheerful man. He has lately been in Holland, and has come back delighted with the people ; they are so like the English, he says, and so cordial and friendly to the English ; he says it is very pleasant to find oneself among people who are so much more kindly disposed to us than the generality of foreigners are. They have much more likeness to English people, he says, than the Americans have.

Met also John Moore, who has just returned from Italy ; he says the coldness of the weather there has been something extraordinary. All the way from Sorrento to Munich, he was " never out of sight of snow." He returned over the Brenner, one of the lowest of the Alpine passes, and found snow on it much below the highest part of the pass.

<hr />

18th May, Wednesday.

Went with Fanny, early (10 to 11) to the Royal Academy, and saw the pictures pleasantly. Of course in an hour one could only pick out a few. Charles Lyell's portrait by *Dickenson* is capital. Lady Bristol, by *Graves*, very pretty, and a fairly good likeness.

* Coleridge.

1870.   Of the " subject " pictures (not portraits or landscapes), the two which pleased me most, in this first
view, were *Millais's* " Boyhood of Raleigh," and
Goodall's " Jochabed," (the mother of Moses) " de-
positing her Infant among the Bulrushes."

There are some very beautiful landscapes, par-
ticularly " Afternoon," by *Creswick*.   " Wind, Rain
and Sunshine at the upper end of Loch Tay," by
Lee.   " A Bright Midsummer Night in Glencoe," by
A. Gilbert; and some admirable sea pieces by
Cooke.

Resumed my work on Hawkwood.

Here is a story told me by Charles Lyell : Huxley
once took his seat in a railway carriage, having a
tortoise on his lap.   A guard came up and said—

" If you please, sir, that animal must be paid for
as a dog."

Huxley remonstrated, and the guard went away
to get further instructions from his superior.

Presently he came back and said, " I find, Sir,
that *cats* is *dogs* and *rabbits* is *dogs*, but a tortoise is
an *insect* and don't pay."

May 19th, Thursday.

Splendid weather, very warm.   Visit from Admiral
Spencer.

Our dear Kate Hervey, came to stay some time
with us.   William Napier dined with us.

May 20th, Friday.

Went again to the Royal Academy.   *Millais's*
" Knight Errant " is a splendid piece of painting ;
the Knight's armour and the trunk of the tree

appear to me admirable in execution. The naked 1870.
damsel is not remarkably beautiful.

Watts' "Fata Morgana" is a fine imaginative
design ; and the half unreal look of the nymph is
appropriate to the idea.

I do not admire *Landseer's* picture of the "Queen
and Prince," but his "Group of Monkeys" (one of
them nursing another which is sick) would be
delightful, if it were not too touching. The third
monkey who is eating an orange with immense
enjoyment, introduces a comic element into the
picture.

Gerôme's "Jerusalem" is a picture which makes
a gloomy and solemn, and almost awful impression.
Jerusalem dimly seen from the hill of Calvary, under
the mysterious supernatural darkness which attended the Crucifixion ; the guards and spectators
departing ; and quite in the foreground the shadows
thrown from the three crosses, which themselves
are not in the picture.

The same *Gerôme's* "Death of Marshal Ney" is
too painful and distressing.

Minnie, Sarah and Edward dined with us.

———

Saturday, 21st May.

Visit from Lady Mary Egerton—agreeable as
ever ; George Napier and Leonard came to
luncheon. Went on with Hawkwood.

———

Sunday, 22nd May.

Charles Lyell and Douglas Galton called on us—
both very pleasant.

1870.   Charles Lyell told me an anecdote of the sagacity of the Falkland Island Seal, in the zoological gardens: a photograph was taken of it, and the keeper took a good deal of trouble to get the animal into the right position. Ever since, whenever the creature sees a photographer passing near with his apparatus, it shuffles to the spot where it was placed to be photographed, and *poses* itself properly.

Charles Lyell tells me that an important geological discovery has lately been made in the Isle of Arran, a classic ground of geology: a discovery of numerous trunks of Sigillaria, upright (that is at right angles to the stratum on which they are based) with their roots preserved, and evidently standing as they grew, and buried in a bed of volcanic ash. And not merely one, but two or three deposits of such trees have been found, all rooted in their original soil, buried in volcanic ashes, and finally tilted up at a high angle. They are almost counterparts of those buried trees which Lyell described in Nova Scotia. And these are new discoveries, it seems, in the Isle of Arran, one of the districts longest known to geologists, and oftenest explored by them.

―――――

May 23rd.

A very pleasant luncheon party: Minnie and Sarah, Minnie Powys, Octavia and Wilhelmina Legge, George Napier, Admiral Spencer and Mr. Bowyer.

We (Kate Hervey included) dined with the Charles Lyells. A small party:—Mrs. Stackhouse Acton, Miss Moore, a Mrs. and Miss King, John

Strutt, Edward, Mr. Dickenson, the artist, Mr. 1870. Hughes and one or two more.

————

Mary came to luncheon with us. Visited old Sir John Bell, my old Cape friend, now (I am told) in his 89th year, yet full of life, spirit and humour. His story that the people of Lowestoft, wishing to obtain the right of returning a member to the House of Commons, offered that if their wish were granted they would change the name of their town to *Bob-Lowestoft*, in honour of the Chancellor of the Exchequer.

Visited Lady Smith and Lady Rich.

————

We went to the flower show at the Botanic Society's Gardens in the Regent's Park. A very beautiful display of flowers as usual.

————

Sir George Young, William Napier and Miss Hoare came to luncheon. Went with Fanny and Kate to a performance of " Glee and Madrigal Union " at St. James' Hall : delightful singing.

————

Much business. We dined with the Rayleighs.

————

We went to an afternoon party at Miss Coutts'— a great crowd. Afterwards called on Lady Adair, who is looking very ill.

1870.   Monday, 30th May.

The 26th anniversary of our happy wedding.—
God be praised.

A pleasant party at luncheon :—Ladies Barbara
and Charlotte Legge, Miss Hervey* (Lord Charles'
daughter), Miss Strutt, Admiral Spencer and George
Napier.   We dined with Minnie.

Tuesday, 31st May.

Finished reading Disraeli's new novel "Lothair:"
this entertained me extremely.

Wednesday June 1st.

A visit from Graham Moore Esmeade.   He is
lately returned from Rome, and he tells me that
there is no doubt that the Pope's mind is weakened.
He is still able to put on a stately, becoming air
and demeanour on great public occasions, but at
other times he is continually bursting into childish
laughter without any reason.

Thursday, June 2nd.

At the Jermyn Street Museum, I met Professor
Ramsay, and with him Dr. Dawson of Montreal, whom
he introduced to me, and whom I had never seen
before, though I had heard so much of him from
Lyell.   He has contributed much to the knowledge
of fossil plants, and is especially distinguished for
his discoveries respecting the fossil flora of the
Devonian or Old Red Rocks of North America :

* Afterwards Mrs. Lock.

indeed, almost all that we know of that earliest 1870.
fossil flora is due to his researches. He told me
that the conditions of the Devonian period in the
most eastern part of America, in Canada and Nova
Scotia, appear to have been much more favourable
to the preservation of the vegetation in a fossil state
than in any other country yet explored; further
west in America (as for instance in Ohio), the
vegetable remains are few and indistinct. But the
traces of plants as yet found in Scotland in the Old
Red, appear, he says, to agree with those which he
has studied in north-east America.

I asked about the age of the sandstones in
Ireland which contain the beautiful fern that
Edward Forbes calls the Cyclopteris Hibernica.
Ramsay thought they were at the bottom of the
Carboniferous series : Dr. Dawson, that they were
uppermost Devonian.

Met Joseph Hooker at the Athenæum. He said
that the exotic trees and shrubs at Kew have
suffered terribly from this last winter, though he
does not think that anything of importance is
absolutely killed. The Sequoia sempervirens at
Kew, as well as with us, has been very severely cut
and browned by the winter. Hooker and I agreed
that the greatest wonder and puzzle in the natural
sciences, in the present state of our knowledge, is
the former existence (which seems indubitable) of a
varied and luxuriant vegetation in the Polar regions.
He is now employed in describing the Nepentheæ
and Rafflesiaceæ for De Candolle's Prodomus.
He has been working at the Rubiaceæ for his own,

1870. and Bentham's Genera Plantarum, and he tells me that he finds an immense number of untenable genera among them:—genera originating in too absolute a reliance on special characters without regard to natural grouping.—Also that *dimorphism* of flowers appears to be very common in that family, and has misled many botanists as to the limits of genera and species.

We dined with the Bishop of Ely.

——————

Saturday, June 4th.

Went to a tea party at Katharine's.

——————

Sunday, June 5th.

We went to early morning service at St. Peter's, Eaton square, and received the Communion. Went again to afternoon service.

——————

Monday, June 6th.

Emily (William) Napier came to see us, just arrived from abroad.

——————

Tuesday, 7th June.

Henry arrived.   Rachel Bruce* dined with us.

——————

Wednesday, 8th June.

Went with Henry and Kate to the Horticultural Society Flower Show, at South Kensington. A very beautiful display of flowers.   We met William

* Afterwards, Mrs. Vernon Harcourt.

Napier.   There was especially, a multitude of 1870. wonderful and beautiful orchids.   Many fine ferns and palms and strange Cycads and Aroids.   I noted particularly — Struthiopteris orientalis, a novelty, but without fertile fronds ; the barren ones much like in general appearance to those of Struthiopteris Germanica.

An immense plant of Asplenium Nidus.   A magnificent plant of Nepenthes Rafflesiana, with multitudes of its great variegated pitchers hanging down on their long thread-like stalks from the ends of the leaves.   Leschenaultia biloba, a species new to me, with large flowers of the most brilliant blue, like the finest blue lobelias.

A fine flowering species of that very curious orchid Uropedium Lindenii, which I first saw at Bentham's Linnean party a few years ago, and made a note of.   Several of the *Alocasias* and *Anthuriums (Aroids)* have wonderfully fine leaves, remarkable for size, colour, gloss and veining ; Alocasia Metallica in particular.

Rachel Bruce and Sydenham Hervey dined with us.   We went to a small evening party at Minnie's.

Lord Lilford told me that the *Pinsapo* fir grows on the Serra da Estrella in Portugal, as well as in the south of Spain.   He says that the proper pronunciation of the name is with the second syllable long or accented

————

Saturday, 11th June.

I see in the newspapers the death of Charles

1870. Dickens. It gives one a shock. I did not know him personally, and I have not admired many of his later writings—in fact I had grown weary of his *manner*, but one cannot be unconcerned at the departure of an original and remarkable genius, who was also a good man.

We had a dinner party—Lady Bell, Harry and Katharine Lyell, the Louis Mallets, the Bowyers, Julia Moore, Minnie Napier, Sally, Mrs. Byrne, George Napier and Edward.

Sunday, June 12th.

After afternoon Church, we drove to Combhurst, dined there, and spent the evening very agreeably with Mr. and Mrs. Smith and their daughters, Mr. Godfrey Lushington, and Mr. William Nicholson.

Monday, June 13th.

Went to the Linnean Society, and looked into the Royal Academy.

Lady Campbell dined with us—charming.

Tuesday, June 14th.

Called on Lady Head.

Our dinner party—Mr. Powys and Lady Mary, the Angersteins, Lady Octavia Legge, Colonel and Mrs. Gardner, George Lock, Minnie, Catty and George Napier, Douglas Galton and Edward. Some others in the evening.

Wednesday, June 15th.

Went out with Fanny and Kate Hervey, first to

the South Kensington Museum, saw the cartoons of 1870.
Raphael, the Vernon Gallery, &c., then to an
afternoon party at Norah Bruce's, then to the
Dowager Lady Lilford's.

Read the reports of some of the speeches in the
House of Lords on the Irish Land Bill.—It was a
fine debate : Lord Salisbury's speech very powerful:
Lord Dufferin's clever and interesting, but to my
mind not at all conclusive as a defence of the bill.

Thursday, June 16th.

After a pleasant visit to Katharine's I went into
the Zoological Gardens and lounged there for an
hour very agreeably. The day being excessively
sultry, the ducks and other water birds appeared
very happy, and indeed rather enviable. In the
aviaries, too, most of the birds were very busy and
lively ; this is always to me a very lively part of
the collection. Many of the birds are very beautiful,
especially the doves, of which there is a great
variety. The horn bills by no means beautiful, but
very curious. The snakes are lively to-day. The
carnivora mostly fast asleep.

The Campbells, William Napier and Edward
dined with us.

Friday, 17th June.

We went with Kate Hervey, Lady Campbell and
her daughter Annie to the British Museum, and
spent some hours there. We met Professor Owen
who looks much aged. He pointed out to us the

1870. enormously large eggs of an extinct bird the
Æpyornis, from Madagascar: these he supposes may
have been the Rok's eggs of the Arabian Nights, as
the Arabians, by their commerce in the Indian seas,
may have received vague reports of these eggs,
and have imagined a bird in proportion to them.
The real Æpyornis, however, seems (as I understood
him) to have been wingless, or nearly so, and utterly
unlike the Raptorical birds. He pointed out also
the skeleton, (one nearly complete) of the Moa, or
Dinornis of New Zealand, which had legs of
enormous size and strength, apparently out of all
proportion to the rest of its frame. He spoke of the
various wingless, or nearly wingless birds, and their
remarkable distribution:—the Moa confined to New
Zealand, the Æpyornis to Madagascar, the Dodo to
Mauritius, the "*Solitaire*" to the island of Rodriguez,
the Cassowary to the Moluccas, and so forth.

It was very interesting and pleasant in this visit
to the Museum, to see Annie Campbell's eager and
intelligent curiosity, her ardour to learn all she
could, her attention to all she saw and heard. She
is just 11 years old, a charming child, one of the
cleverest I know.

———

19th June. Sunday.

We called on Lady Rodney. In the evening we
read a fine sermon by Dr. Temple, on Resignation.

———

Monday, 20th June.

Our dinner party:—Charles and Mary Lyell,

Lady Head, Sir Edward Ryan, Mr. and Mrs. 1870.
Gambier Parry, Mr. Fuller Maitland, Minnie
Napier, William Napier, Edward, Mr. and Mrs.
Longley—all pleasant. Mary Lyell told me that
Dr. Bigelow of Boston, who is 83 years of age,
lately made an excursion with his family to
California, and they went from Boston to San
Francisco, across the whole breadth of North
America in 9 days, by the lately finished Pacific
Railway : and that without any particular hurry.

———————

Wednesday, June 22nd.

Mr. Clark (of Trinity College Cambrige) came to
luncheon with us.

Read with much approbation, Mr. Forster's speech
in the House of Commons on the Education Bill in
the debate of the night before last. I heartily
approve of the course he has adopted, of determined
opposition to those who wish to exclude religion
from schools. Again went to the National Gallery.
We dined at Lord Belper's—met Miss Stanley, Mr.
Mundella, Douglas Galton, Mr. Wedgwood and
some others.—Afterwards Fanny went to a concert
at Mrs. Peploe's.

———————

Thursday, 23rd June.

The reading of the Parliamentary debates now
take up much of one's time, as the discussions on
the Education Bill are of great interest and importance. I have read Gladstone's speech winding
up the debate, with great satisfaction ; Lord
Sandon's also, and Mr. Baine's and Henry Bruce's.

1870. There is much also that I approve and admire in
Mr. Mundella's, though I differ from him on some
points.

Drove out with Fanny and Kate, and left cards
at Holland House.

We dined with Minnie—met Mr. and Mrs.
Fazakerly, Minnie Powys, Mrs. Ellice and Cissy,
Mr. Goodlake, George and Edward.

* * *

News of the death of Lord Clarendon ; a great
public loss.   His loss will, I think, be especially felt
by the present Ministry, not only from his long
experience, and great knowledge of foreign affairs,
but as a sobering and steadying element in a Cabinet
which inclines too much to Radicalism.

Visited the South Kensington Museum.   John
Hervey came to stay with us, and his cousin, Lord
John, also dined with us.

* * *

Fanny and Kate went to the House of Commons.
I went to Mrs. Evans Lombe's afternoon party.

We went to a pleasant evening party at the
Charles Lyells.   Many whom I know and like there,
and not too crowded.

Sir Edward Ryan said of the Irish Land Bill,
that it is justifiable only on the assumption that the
Irish are in a state so completely anomalous and
excptionable, that the rules which are applicable to
all civilized communities do not apply to them, and

they require an entirely exceptional legislation. "It 1870.
"is only on such grounds that it can be justified
"for it is quite wrong in principle."

It must be regarded as an *Irish* measure, and
therefore an exception to all rules. This coincides
most exactly with my own opinions upon the Bill.

Sir Edward Ryan thinks there will be much
difficulty in filling the place of Lord Clarendon.
The Foreign Office is, he thinks, the most difficult,
complicated and laborious of all the departments of
Government. I should have thought the Home
Office more difficult, but he thinks *not*.*

I learn from Leonard Lyell, that Æpyornis of
Madagascar (see June 17th) is believed to have
been allied to the Deinornis, and from the analogy
of that and other allied forms, it is thought probable
that the bird was not nearly so large in proportion
to the egg, as would be inferred from the general
proportions of birds.

_____

June 29th, Wednesday.

Admiral Spencer and Mr. Lott came to luncheon.
We went to afternoon tea at Lady Lilford's.

Lady Rayleigh, Miss Strutt and John and Charles
Strutt, dined with us.

_____

Thursday, 30th June.

Arthur Hervey came in to see us for a few minutes

* Note (12th September, 1870).—This anticipation of the difficulty of filling
Lord Clarendon's place has been amply fulfilled, and he has been sorely
missed in the tremendous crisis which has since occurred.

1870. in the afternoon.   We dined with the Angersteins—
I sat by Mr. Inglefield, a pleasant person.   There
were also the Gardners, Sir Roderick Murchison,
Sir Fenwick Williams, &c.

————

Friday, 1st July.

Our dear friend, Arthur Hervey (Bishop of Bath
and Wells) dined with us;   as delightful as ever,
and looking remarkably well.   He is indeed the
youngest looking man of his age (61) that I know.

————

Saturday, 2nd July.

Drove out with Fanny.

We looked into the Horticultural Gardens, and
called on Mrs. Greig.*   We, including Kate Hervey,
dined with the William Napiers at Wimbledon—a
pleasant little party.   There was Sir Bartle Frere,
so much distinguished for his administrative ability
and benevolence in Scinde, and Western India, and
famous also as a geographer, a very interesting man,
peculiarly pleasing in his manners and agreeable in
his conversation.

He told me much about the physical geography
of the countries near the Indus.

The rest of the party besides the family, were
Mr. Goldschmidt, the husband of Jenny Lind, and
a Mr. Harding.

Sir Bartle Frere was, as I learn from Murchison's
anniversary address of this year, one of the original
founders of the Geographical Society.

---

* Mr. Greig, son of Mrs. Somerville, by her first husband.

Sunday, 3rd July.

A visit from Arthur Hervey. We went to afternoon Church at St. Peter's, Eaton Square. Read to
the servants, a good sermon by Dr. Temple.

———

Monday, 4th July.

I took Kate Hervey to the Athenæum. Edward
met us there and showed us over the club.

Fanny and I went to stay with the Charles
Baring Youngs, at Oakhill, near Barnet. Met there,
Herman Merivale. The house and gardens very
pretty.

———

Tuesday, 5th July.

We returned from Oakhill to Eaton Place. We
had a pleasant dinner party, consisting of Mr. and
Mrs. Bentham, Mr. Boxall, the Carrick Moores,
Charles Mallet, Mrs. Lombe, Mrs. and Miss Ellice,
Patrick Blake, Minnie Napier, and Edward.

———

Wednesday, 6th July.

We went to a garden party at Holland House—
a very pretty scene.

———

Thursday, 7th July.

We had a very pleasant dinner party—Arthur
Hervey, the Bishop of Ely, and Mrs. Browne, Sir
Bartle Frere, the Charles Lyells, Mr. and Mrs.
Mills, Mr. Walrond, Philip and Pamela Miles,
Rachel Bruce, Miss Stanley, Mr. William Gurdon,
besides Kate and John Hervey, who were staying
with us.

1870.   Sir Bartle Frere said that our knowledge of Central Asia has made great progress in these latter years, and has advanced at a much more rapid rate than formerly ; those countries are found to be not so inacessible* as they were long supposed. What the rulers, and the people too, of those countries have the greatest dread and horror of (he said) are spies—anything that excites their suspicions of dark and secret designs ; they are by no means so hostile to solitary travellers who go openly forward on their particular pursuits without disguise or concealment.

Talking of badgers, (Arthur Hervey having mentioned that the badger still exists within a few miles of Wells) Sir Bartle Frere said that the common black bear of India is in some measure intermediate in form and habits between the true bears and the badgers. I understood him to say that it is a burrowing animal like the latter. This is the bear, he says, which is described by Williamson as so terribly ferocious, yet it lives more generally on vegetable than on animal food, and is particularly fond of sugar cane.

Friday 8th July.

Fanny and Kate went to Lord's Cricket ground.† I went to Harley street to see Charles Lyell, and walked with him as far as Murrays.‡   I went to see

---

* So absolutely closed against foreigners.

+ The Eton Cricket Match.

‡ John Murray, the publisher.

him again later in the afternoon and had a good talk   1870.
with him.

William and Emily Napier dined with us.

Charles Lyell has received some long letters from
Professor Heer, of Zurich, earnestly contending
against the opinions of Bentham, who (in his Linnean
Society address) has pointed out the uncertainty
of many of the determinations of fossil plants.
The dispute is specially about the Proteaceæ; Heer
eagerly contending that there is evidence of their
abundant existence in Europe, in tertiary times,
while Bentham holds that the evidence has no
validity.

Lyell, who has almost implicit faith in Heer,
eagerly adopts his view of the matter; I, on the
other hand, have great confidence in the sagacity,
knowledge and judgment of Bentham, and think
that he is more likely to be right. But Lyell is
probably quite right in one point—that much of
what we do know concerning tertiary plants, is due in
the first instance to sagacious and happy conjectures
or guesses by Heer and others; and that it is
unadvisable to discourage a mode of enquiry which
has led to such results, although it may have led to
some errors.

_____

Saturday, July 9th.

Dear Kate Hervey left us, to return to Wells
with her father and her brother John. I need not
say we were sorry to part with her. She has been
a delightful inmate of our house ever since the 19th
of May.

We returned yesterday evening from a very
pleasant visit to Lady Mary Egerton at Mountfield
in Sussex.   We went down on Saturday the 9th,
by the South Eastern Railway to Robert's Bridge,
and thence in a small carriage three miles further
(almost continuously uphill) to Mountfield.

Lady Mary's house stands very high, almost on
the top of a great hill, commanding extensive and
delightful views of a beautiful country, hilly and
richly wooded and finely varied—part of the Weald.

Battle Abbey is about four miles off,* and on the
top of a high hill in the distance, we see Fairlight
Church, above Hastings.   The soil mostly sandy,
or a mixture of sand and clay, in parts very clayey;
belonging to the Hastings sand.   It is a country
rich in trees ; near the house are groups of magnificent limes, and at a little distance very fine beeches,
chesnuts and Scotch firs ; in particular an avenue
of noble chesnut trees, which, according to tradition,
are of the same age as those in Greenwich Park.
Hop grounds are the distinguishing feature of
the agriculture, and a very beautiful one.

The oak appears to be the prevailing tree, but
I did not see any particularly large ones.   Young
Mr. Egerton told me that a former proprietor, from
whom his father bought the estate, had cut down all
the oak trees that were of any value as timber.
He also told me that the *hornbeam* is one of the
commonest wild trees or shrubs of the district,—
"one of the weeds of the country," he said.

* Not in sight, however.

Between the house and Mountfield is a deep, 1870.
hollow lane like a Devonshire lane—very pretty;
the steep bank on one side, beautifully clothed with
luxuriant ferns. It affords a characteristic though
small section of the Hastings sandstone beds of
hard, sandstone rock, amidst many of loose
crumbling sand.

What above all made our visit so pleasant, were
the agreeable and estimable qualities of the family;
for of course (Mr. Egerton's death being so recent)
it was strictly a family party.

Lady Mary appears to me a very superior
woman : of a very strong and active mind, highly
cultivated, with a remarkable turn for mathematical
and physical science : at the same time with great
good sense, and quite free from the unfeminine
hardness so often remarked in " strong-minded "
women; her manners very gentle and pleasing.
Her daughters (four, of whom two are " out ") are
very agreeable and interesting girls, clever, lively,
and well-informed, and the eldest son also made a
favourable impression on us.

On the 11th, Lady Mary took us in her carriage
to Hawkhurst, where we visited the Herschells, and
spent an hour very pleasantly with them.

Sir John is looking very old in the face, but seems
not very infirm in body, and quite active and clear
and animated in mind : he received us very
cordially, and talked of old Cape days with much
zest. He showed me a fine spike of Disa cornuta,
which he had preserved in a peculiar preparation
ever since he was at the Cape. This process has

1870. perfectly preserved the form and structure, but the colour has totally disappeared. Sir John told me of a curious variation or "monstrosity" which he had lately observed in a wild specimen of the common Foxglove: some of the flowers having a deep pouch —almost a spur, projecting from the under-side of the corolla. He showed us a plant of the Birds-nest Orchis, which his daughters had found in a neighbouring wood and removed to their garden. Lady Herschell was looking very well, and was (as were her daughters also) very friendly and agreeable.

From Hawkhurst we returned to London by railway from Etchingham.

On Tuesday, the 12th, Minnie Napier came to luncheon. After it, I went with Fanny to call on Lady Georgina Bathurst,—afterwards to an exhibition of Raphael's drawings.

We had visits from Charles and Mary and Edward.

Wednesday, 13th July.

Sir John Kennaway came to luncheon. Edward dined with us. Minnie and Sarah came in the evening.

Thursday, 14th July.

Emily and William Napier came to luncheon with us. We visited the Charles Lyells. Fanny bought some Java sparrows.

Friday, 15th July.

We drove to Battersea Park, and saw the sub-tropical garden. We returned to Barton, arriving 1870. safely at our dear home—thank God.

———

A violent thunderstorm this morning; afterwards a beautiful and very hot day. There has been rain enough lately to counteract, in a great degree, the effects of the long drought in the spring and early summer, and I find the lawns and pastures here (to my surprise), beautifully fresh and green. The foliage of the trees is very rich and luxuriant, and the flowers fine: roses and strawberries have been in great perfection this year. The hay crop, owing to the long drought, has been a complete failure, but the rain has come in time to save the turnips and other green crops, and the corn I am told, (at least in this parish) promises very well.

I do not find that many of our exotic trees or shrubs have suffered materially from the severe winter. The Benthamia, indeed, is nearly killed : and our largest Araucaria (planted by my father in 1832) is quite dead : but this last has never been really healthy since 1860. Sequoia sempervirens, which had been turned entirely brown, as if burnt, has recovered. Æsculus Indica is in a very flourishing state, and has flowered plentifully : its blossoms are passing off. The great Catalpa has a very great promise of blossom.

The news is most serious, indeed, awful. France has declared war against Prussia. As far as I can learn or understand at present, it appears to me one of

1870. the most causeless, unjustifiable, wicked wars recorded in the modern history of Europe.

Monday, July 18th.

We dined at the Milbanks.

LETTER.

Barton,
July 21st, 1870.

My Dear Katharine,

    This war is very shocking, and seems to have burst upon us so suddenly—like a tropical storm, that it is almost bewildering. With all the new inventions in the science of destruction, the slaughter is likely to be frightful, and though they say that the war will be short, who knows? Who can tell where it will stop, or what other nations may not be dragged into it?· I hope *we* shall manage to keep out of it; I should think that our present Government really wish for peace, but there is no knowing what "complications" may arise, which may make it necessary for us to choose between war and disgrace. I believe indeed, that, thanks to the great deeds of the present Ministry in the way of parsimony and retrenchment, we should find ourselves totally unprepared for war. The feeling in this country seems to be very generally in favour of the Prussians, and I should think it is to a great degree right, though I believe that in most quarrels, both are in some degree to blame. But it appears to me that the French, or at least a great party in 1870. France, have been meditating war and preparing for war ever since '66, and that now, believing themselves to be prepared, they have eagerly seized the first pretext.

This place is in great beauty, and very enjoyable in this glorious weather: the grass very green, the trees in fine foliage, and flowers in abundance. I hope it will be looking equally well when you come. The ferns are flourishing.

Cecilia and her children (three of them that is), arrived yesterday, and it is a great pleasure to have them here. Fanny has probably told you of the sad event that has happened at Bury since we returned home—the death of Mrs. Bevan. She was a handsome, intelligent, cultivated, agreeable woman, and will be much missed in the society of the neighbourhood, and it makes one very sad to think of the loss to her poor husband.

I have begun to examine and arrange the dried plants from Trinidad, which Kingsley gave me.

> Believe me ever,
> Your very affectionate brother,
> CHARLES J. F. BUNBURY.

JOURNAL.

Sunday, 24th July.

We all went to morning Church.

Read a sermon on the War, preached in London last Sunday by Mr. Stopford Brooke, and now

1870. published. It is most eloquent and powerful, very noble in its moral tone and in its denunciation of the wickedness of such a war as this ; but rather injudicious and irritating in the tone in which it speaks of the French, and too extreme, I think, in throwing too positively on them the whole guilt of the war.

————

Thursday, 28th July.

Lady Cullum and Miss Birch and the Abrahams dined with us.   Mrs. Abraham and Miss Birch sang delightfully.

The large Catalpa tree in the arboretum is now in glorioŭs beauty,—one mass of blossoms.

The War naturally occupies much of one's thoughts.   No serious engagement has yet taken place, but it is evident that both parties are thoroughly in earnest, thoroughly determined, and that the struggle will be a tremendous one.   The enthusiasm in France is described as great, and on the other hand, the earnest German feeling—not merely Prussian—seems to be thoroughly roused.   For my own part, I wish success to the Germans.   I think their civilization is of a type more congenial to ours, and their spirit less dangerous to the rest of Europe than that of the French ; and in this particular war, I think the French more to blame.

In fact, I have no doubt that the French, or at least the Emperor and a strong party in France, have been designing war and preparing for war ever since 1866 ; and now that their preparations are complete, they seize the first pretext.

The French government make great professions of 1870. friendship towards us, and of their good intentions of respecting the neutrality of Belgium ; but one can hardly feel very secure. I fear it is not unlikely that we may before very long, be placed in a situation in which we shall have to choose between war and a disgraceful desertion of our allies. Edward believes, from what he has heard from various quarters in London, that there is a strong feeling, even among the lower classes (of course not the lowest), that if Belgium be attacked, we *must* fight for her. This is as it should be. But are we prepared for war ? Edward is satisfied, from all he heard in London, that our navy is very efficient, much superior to that of any other nation whatever ; and much stronger than it has ever been since the great war. But our army is very deficient in numbers at any rate. The Ministry, who have for two years past been so unmercifully cutting down our military and naval establishments, and so eager to sacrifice everything to economy, now see the necessity of retracing their steps, and are buying back at enormously increased rates, the stores which they lately were in such a hurry to sell.

Friday, 29th July.

Scott (who is lately returned from Mildenhall), tells me that the crops of wheat and oats there are very good, except on the very light sandy lands ; and on the so-called skirt lands (bordering on the fen), uncommonly good. On what are usually con-

1870. sidered the best soils in the parish, such as the
Wammell farm, the crops are not so good, relatively,
as on some of the other lands ; for where the chalk
is near the surface, the soil is very liable to be
too much burnt in such a season as this. The
cutting of wheat has begun pretty generally, he tells
me, in that parish ; the fen district of course ex-
cepted.

———

Monday, 1st August.

We went to a garden party at the Abrahams, and
spent the afternoon there : met most of our neighbours ; pretty singing.

The harvest seems to be now pretty generally
begun in this neighbourhood, and with very fair
prospects. In driving to Risby this afternoon, I
observed much wheat cutting. At Mildenhall,
Scott tells me, a good deal of wheat was carried,
and in excellent condition, by Saturday last, the
30th.

———

Tuesday, 2nd August.

I find there is much difference of opinion on the
question, whether France or Prussia is most to
blame in the origin of this war : and no doubt there
is room for very different views. My own
impression is, that France, having been irritated
and mortified by the great successes and brilliant
military achievements of Prussia, and also by the
haughty tone of Prussia, in 1866, and by the
apprehension that she (France) might be suspected

to be no longer the greatest military power in 1870. Europe, has ever since been in a state of sensitive irritability, which made her eager to find cause of quarrel in the first shadow. On the other hand, the Prussian Government, haughty and ambitious and confident in their strength, were probably not at all anxious to avoid giving cause of offence, and if they did not seek war, were not very desirous to avoid it. The case was like that of some duels in the old duelling days, when two men would go on nourishing and nursing irritation and ill-will against each other, till any trifle sufficed to bring on a fight. Our Government seems not to have been wanting in endeavour to mediate, but in vain.

I have just received a letter from Joanna Horner from Paris; she and Susan arrived there yesterday morning (August the 1st), having left San Marcello (in the Apennines, near Pistoia) in the afternoon of Saturday the 30th. She says that they have made out their journey "capitally," with "no interruptions, "no passports required for ladies, perfect civility "and attention, no signs of war, except at Dijon, "where we stopped an hour to allow a train of "horses to pass." They came by the new railway over Mont Cenis, and by Chambery, Culoz, Macon, to Dijon.

————

Wednesday, 3rd August.

Henry and his son Henry arrived.

1870.

Thursday, 4th August.

Engaged all day at Bury on the grand Jury at the Assizes—many serious cases—discharged at 6 p.m.

Dear Susan and Joanna arrived, looking very well and as agreeable as ever. A very pleasant evening.

News of fighting between the French and German armies: the former have taken possession of Saarbrück, after driving out the Prussian force which occupied it: but the telegraphic accounts are so contradictory, that it seems helpless to arrive at any safe conclusion, either as to the numbers engaged, or as to the loss on either side.

The name of Saarbruck is familiar to me in a very different relation; it is a noted locality for fine fossil plants of the coal formation.

Friday, 5th August.

News of another action, and seemingly a more considerable one, but as yet we have only the Prussian official report, and the telegraphic accounts are never very clear. It appears that the Germans (by their own account) attacked and carried the encampment of a French Division at Weissenburg; that a French General Officer, General Douay was killed, and 500 unwounded French prisoners taken.

Katharine, with her husband and Arthur and Rosamond, and Leonora with her two daughters, arrived in the evening.

Mrs. Praed and Lady Young and the Miss Waddingtons came to luncheon. Afterwards we had a large party, chiefly of boys and girls, who spent all the afternoon in various outdoor amusements.

It seems certain that the French had the worst of it at Weissenburg. Their own account (telegraphic) is that three of their regiments of infantry, and a brigade of light cavalry, altogether 7,000 or 8,000 men, were surprised early in the morning by a very superior force of Prussians, and after resisting vigorously for some hours, retreated in good order..

Marshal Macmahon was concentrating his forces in that quarter, and a great battle was expected immediately.

Monday, August 8th.

Important and exciting news. The Prussian army, under the Crown Prince, has gained a great victory over Marshal Macmahon's corps d'Armée (aided, it is said, by portions of two other corps), at a place called Woerth, a little way S. W. of Weissenburg and N.W. of Haguenau.

The Prussians, it is said, have taken 30 guns, six of the new *mitrailleuses*, and two eagles, besides some thousands of prisoners. The number of killed or wounded on either side is not mentioned in the telegrams, but was probably heavy. On the same day, it seems, another portion of the French Army was defeated by another corps of Prussians, near Saarbrück.

1870.   It is probable that the French, though they
began the war, were not really ready, and that the
Germans were able to concentrate their forces more
expeditiously.   There is however no reason to suppose that this victory is at all decisive, and I
hope that the French will recover their ground.
Not that I wish complete success to the French,
either.   I should be very sorry if either party
suffered a crushing defeat or gained an overpowering
victory ;   very sorry if either gained or lost any
considerable territory by the upshot of the war.   In
the meantime it is terrible to think of the amount
of suffering and misery implied in such great battles
as these seem to have been.

Thursday, 11th August.

Fanny, Cissy, and young Henry, Mr. Eddis and
Susan, dined at Hardwick.

Friday, 12th August.

Mr. Eddis completed an admirable drawing of
young Henry.

I looked over some dried ferns with Katharine.
Lady Cullum dined with us.

We are still without any clear and distinct
account of the battles lost by the French near
Weissenburg and Saarbrück on the 6th.   It seems
certain that Macmahon with one corps only, or part
of one corps, was opposed to a very superior
Prussian force.   The French accounts say that

he had only 35,000 men against 120,000 ; and 1870.
it seems difficult to understand, without supposing
great faults in the French generalship, how he came
to be opposed to such odds, as the total force of the
French is certainly not inferior to that of the
Germans. Some of the French telegraphic accounts
imply (as far as they are intelligible at all), that
Macmahon *stumbled* upon the Crown Prince's army
withont being aware that it was so near ; a strange
thing, surely, to happen to a famous General of such
a military nation as the French. However, the
French corps appear to be now concentrated for the
defence of Metz, and the Prussians have not made
any important movement in advance within the last
few days. In the meantime, there has been tremendous excitement in Paris, but as *yet* no revolution, only a change of Ministry.

———

Sunday, 14th August.

At last Macmahon's official report of the battle
near Woerth is published, but it is very hard to
understand, especially without topographical plans or
maps on a large scale. It does not at all explain, as
far as I can see, *how* he came to be opposed to such
superior numbers ; but it seems (though this is not
quite clear), to represent him as the assailed, not
the assailing party ; contrary to what I had previously supposed. The letters in the newspapers
represent the engagement as unexpected on *both*
sides. This, if true, was probably owing to the
nature of the country (between the Rhine and the

1870. Vosges), which is described as very much broken, hilly, and thickly wooded. There seems to be no doubt that the fighting was desperate, and the loss on both sides very heavy.

— — — —

Went with Susan into Bury, and visited Mr. Borton, by invitation, to see a fine Stanhopia which is in flower in his hot-house. The flowers are beautiful and very curious, looking singularly as if they had been carved in ivory : exactly of the colour of ivory, but thickly spotted over with small circular *eyes* or annular spots of a light-brownish red.

Joanna sang very agreeably in the evening.

— — — —

Katharine and her husband and children went away. Emily Napier and Mr. Powys and Lady Mary arrived. I went into Bury in the morning to a meeting for the election of a Turnkey to the Gaol, and again in the afternoon to a meeting of the Committee about a picture of Lord Arthur Hervey.

— — —

Our dinner party consisting of Lady Cullum, the Hortons, Lord John Hervey, the Ellices, and Patrick Blake.

— — —

More slaughter. Another great and murderous

battle fought on the 16th, between Metz and 1870.
Verdun. The accounts (telegraphic) are, of course,
imperfect, and both parties claim the victory: but
there seems little room to doubt, that, as the result
of the whole, Bazaine's corps has been baffled in its
movement of retreat from Metz upon Châlons, and
that the Prussian armies have interposed between
those two places, and between Bazaine's army and
that which is concentrating at Châlons. I fear that
the prospects of the campaign look very ill for the
French. I say, I *fear*,—not because I feel any
particular zeal for the French, at the beginning of
the war my sympathies were more with the other
side: but I think it very unfortunate that either
party should gain a complete triumph, and so
absolute an ascendancy as the Prussians seem likely
to gain. Besides, I cannot help feeling rather sorry
for the French, and even for the Emperor.

I sent a contribution to the Society for the aid of
the wounded.

We went, a large party, to a garden party at the
Rodwells, at Ampton:—the grounds, some of the
prettiest in this neighbourhood, and the trees very
fine. In particular, noble limes, very fine tulip
trees, and a remarkable number of cedars of great
size and beauty. Mr. Rodwell tells me, he believes
that these cedars were planted by Mr. Calthorpe,
the father of the late Lord Calthorpe, something
more than 100 years ago. There is also a deciduous
Cypress, bearing cones, and a Catalpa larger in the
trunk than ours, but with, by no means, so fine a
head.

August 20th   Saturday.

Dear Henry and Cecilia, with their four lovely
children, went away : needless to say how sorry we
were to part with them.   Emily and young Henry
are quite charming.   It was a sorrowful parting.

Mr. Milbank and Colonel Morris, Lord John
Hervey and the Phippses came to luncheon, and
spent the afternoon with us.   The Powyses went
away.

Sunday, August 21st.

Very cold, though very dry and clear.   The first
day of fires in the evening this summer.

The newspapers at last give us some distinct
accounts of the terrible battles (on the 6th) at
Woerth and at Spicheren, near Saarbrück : official
or semi-official accounts, published by the Prussian
Staff, as distinct, perhaps as they can be without
detailed maps or plans of the ground.   But it still
seems unexplained *how* or why the Prussians were
victorious : for they attacked the French (in each
case) in a very strong position, on steep hills : so
that unless they had a great superiority of numbers
(which is uncertain), the result is hard to understand.   By their enemy's account, the French
fought hard and well for a long time, but when at
last it became necessary to retreat, the retreat soon
degenerated into a rout and a panic flight.   The
total amount of slaughter, in all these numerous and
hard fought battles, must be perfectly frightful.   It
seems that, besides the two on the 14th and 16th,

there was another great battle on the 18th, at a 1870.
place called Rezonville, near Metz: the event
variously represented.

———

Monday, 22nd August.

Minnie and Theresa Boileau arrived.—Very glad
to see them.

The harvest is now finished in this parish, and
the weather has been very favourable for getting it
in.   On Loft's farm, which is the heaviest land
(most decidedly clayey soil in the parish), the crops
are very good: also on the Home farm, which is
heavy land too, and formerly damp.   On Cooper's
farm, which is generally considered the best land in
the parish, crops below the average, both wheat and
barley.   On Paine's farm, also light.   On the
Vicarage farm, good.   On Phillips' (the Burý field)
farm, barley good, wheat not so good.—(From
information of Scott).

———

Tuesday, 23rd August.

We two, with Minnie Boileau and Leonora, went
in the open carriage to call on the Milbanks, at
Ashfield, and saw their beautiful garden.   Called
also on Sir Henry Blake.

———

LETTER.

Barton,
Bury St. Edmund's
August 26th, 1870.

My Dear Edward,

      I was extremely glad to receive this morn-

1870. ing your letter from Neuchâtel, for I was becoming uneasy at not hearing from you, and feared you might have met with some detention or some disagreeable adventure, as the French appear (by the newspaper accounts) to be in a state of excitement and suspicion which might easily become dangerous. I am very glad indeed that you made out your journey comfortably, and are well and enjoying the beauty and tranquility of Switzerland, undisturbed by the horrid passions of man.

What a bewildering succession of great and terrible events there has been since yon left us! and how different the course of things has been from what was expected. I most heartily and thoroughly agree with all you say on the subject. The amount of slaughter, already, is perfectly frightful, and the amount of sorrow and suffering and misery of every kind, which that slaughter implies, makes one sick to think of. What a weight of guilt rests on the souls of the two men you allude to. My wishes have rather changed since the beginning of the war : I now feel for the beaten party, and wish that the others may receive a check ; but one does not see much probability of it. I do hope that Paris will not be besieged, it would be too horrible.

Henry and Cissy and their children left us on the 20th ; Emily Napier (*single* Emily) had arrived on the 17th : and our party at present in the house, consists of *her*, Leonora and her two girls, Susan and Joanna and two Miss Boileaus : so I am the only gentlemen to *nine* of the feminine gender. Next week we shall be alone with the three sisters.

I am reading the 2nd volume of Kaye's "Sepoy 1870.
War," and the 6th of Burton's " History of Scotland."

With love from Fanny,
Believe me ever,
Your very affectionate brother,
CHARLES J. F. BUNBURY.

JOURNAL.

Saturday, 27th August.

The particulars of the war, and especially of the
great battles near Metz, which are now given in the
newspapers, are *terribly* interesting.   Judging from
these accounts, they must have been some of the
most sanguinary battles in modern times.   The
Prussians themselves estimate their own loss in
killed and wounded since the beginning of the
campaign, at fully 50,000; and it is probable that
the French has not been much less.   What a
weight of guilt rests on the souls of the men who
caused the war!   And it is by no means over yet;
for the French, in spite of all their disasters, seem
thoroughly resolute, and determined even to stand a
siege of Paris, rather than submit to a humiliating
peace.

Monday, 29th August.

I went with Susan to an afternoon garden party at
John Phillips'.   Good company: pretty garden.
Saw my grandfather's caricatures.

Leonora and her two dear girls went away.

It is shocking to read of the miseries caused to the unfortunate people of Strasbourg by the siege and bombardment now going on : lamentable to think of the ruin of that interesting old city, and above all of the (too probable) destruction of that glorious cathedral. I fear too that the interesting Museum of Natural History may be destroyed, and Professor Schimper's life must be in great danger. A siege of Paris would be a calamity to all European civilization, yet I fear it is very probable.

_____

September 1st, Thursday.

Fanny and Susan went to Mildenhall, and returned late.

_____

September 2nd, Friday.

Yesterday I received a most interesting and remarkable letter from Charles Kingsley. He says, speaking of the War :—

"I confess to you that were I a German, I should " feel it my duty to my country to send my last son, " my last shilling, and after all, my own self, to the " war, to get that done which must be done, and " done so that it shall never need doing again.

" I trust that I should be able to put vengeance " out of my heart ; to forget all that Germany has " suffered, for 200 years past, from that vain, greedy, " restless, wicked nation ; — all even which she " suffered—women as well as men—in the late

" French war ; (he seems to mean the great war in 1870.
" the first Napoleon's time) :—though the Germans
" do not forget it, and some of them ought not.
" But the average German has a right to say :
" '*Property, life, freedom, have been insecure in Germany*
"*for 200 years, because she has been divided. The*
"*French kings have always tried to keep her divided,*
"*that they might make her the puppet of their ambition.*
"*Since the French Revolution, the French people (all of*
"*them who think and act, viz., the army and the*
"*educated classes), have been doing the same. They*
"*shall do so no longer. We will make it impossible for*
"*her to interfere in the internal affairs of Germany.*
"*We will make it an offence on her part —* after
" Alfred de Musset's brutal song—*to mention the very*
"*name of the Rhine.*' As for the present war, it
" was inevitable, soon or late. The French longed
" for it.   They wanted to revenge 1813-1815 ;
" ignoring the fact that Germany was then avenging
" —and very gently—1807.   Bunsen used to say to
" me—I have seen the tears in his eyes as he said
" it—that the war must come, that he only (*two or*
"*three words are illegible to me*) prayed that it might
" not come till Germany was prepared, and had
" recovered from the catastrophe of the great
" French war.   It has come, and Germany is
" prepared—and I would that the good old man
" were alive, to see the battle of Armageddon, as he
" called it, fought, not as he feared, on German,
" but on French soil.

" As for this being a *dynastic* war, as certain
" foolish working men are saying, who have got still

1870. " in their heads, the worn out theory that only kings
" ever go to war,—it is untrue.   It is not dynastic
" on the part of Germany.   It is the rising of a people
" from the highest to the lowest, who are deter-
" mined to be a people, in a deeper sense than any
" republican democrat, French or English, ever
" understood that word.   It is not dynastic on the
" part of France."

———

Sunday, 4th September.

We went to morning Church ; a beautiful sermon
from Mr. Percy Smith, in aid of the fund for the
wounded.   Walked about the grounds with Fanny,
Susan and Joanna.

———

Monday, 5th September.

The events which have crowded upon us in the
last few days are perfectly astounding.   The
succession of desperate battles near Sedan, the
total defeat and capitulation of Macmahon's army,
the surrender of the Emperor ; and now as it seems,
a revolution in Paris (though our information about
this is still incomplete), one can hardly persuade
one's self that all these things are real.   Many
wonderful things have happened in my time, but
nothing so wonderful, I think, has happened since
1815, as this collapse of the second French Empire,
and of the power of France.   And what is to follow
remains still quite in the dark.   Will peace be the·
consequence ? not yet, I fear ; for the French after
all their losses, appear to be more exasperated than

depressed, and not yet at all inclined to submit. 1870.
Nor do the Germans seem likely to be moderate or
forbearing in their use of victory.

Tuesday, 6th September.

There has indeed been a revolution at Paris, the
Emperor deposed, a Republic proclaimed, and a
Provisional Government appointed; and all without
struggle, violence or bloodshed. Truly these are
wonderful times. I heartily rejoice to read that the
Empress and Prince Imperial are safe out of
France.

September 7th, Wednesday.

Went up to London.

Thursday, September 8th.

Received another interesting letter from Kingsley.
He says :—

"Since Waterloo, there has been no such event
"in Europe. I await with awe the Parisian news of
"the next few days. As for the Emperor: while
"others were bowing down to him, I never shrank,
"even to the highest personages, in expressing my
"utter contempt of him and his policy. It is now
"judged, and he with it, by fact, which is 'the
"voice of God revealed in things,' as Bacon says.
"And I, at least, instead of joining the crowd of
"curs, who worry where they lately fawned, shall
"never more say a harsh word against him. Let
"the condemned die in peace, if possible : and
"he will not, I hear, live many months, perhaps not
"many days. Why should he wish to live ? This

1870. "very surrender may be the not undignified farewell
"to life of one who knows himself at his last."

Called on M. and Mme. de Tourgueneff,* at the
Hyde Park Hotel.

William and Emily Napier dined with us.

———

Friday. 8th September.

Madlle. de Tourgueneff and her brother dined
with us.

LETTER.

48, Eaton Place, S. W.
September 9th, 1870.

My Dear Katharine,

Very many thanks for your delightful letter
from Ambleside, which has interested me very
much. I had been impatient to hear from you, and
am delighted that you have been spending your
time so agreeably. I well remember that beautiful
and charming country, and can well imagine your
enjoyment of it, especially as you have had the
advantage of fine weather, which is not very usual
there. I quite enter into your feeling of regret at
not being able to climb the hills as your children
can do: the young have a great advantage there;
I daresay that if I were of the party, I should find
that I could not do what I could in '44: yet I hope
to get a little into practice of walking among the
Mendip hills. I found about Ambleside in '44, I
remember, all the plants you mention except the
Lobelia,—the Meconopsis I found on the shore of

* They had escaped from Paris.

the lake near Bowness, and Hymenophyllum 1870.
Wilsoni about the pretty waterfall called Stockgill
Force, together with Bartramia Halleriana. I well
remember the beautiful profusion of Cryptogramme
crispa on the mountains, and particularly in the
pass between Ambleside and Keswick. You do not
mention whether you have got any ferns altogether
new (as British species) to your collection : but this,
indeed, is hardly likely.

What a delightful time your children must have
had in the Lake country.

What wonderful times we live in! What a rapid
succession of strange and terrible events in the last
six weeks,—and what may yet be to come, in the
next month or week, is beyond any one's power to
calculate.

I am glad your good old aunt has (though late)
made up her mind to escape from Paris, and not to
stand the risk of a siege. Fanny has just had a
letter from her from Boulogne, where she has safely
arrived, and is waiting for calm weather to cross
over.

I, yesterday, saw poor M. and Madame de
Tourgueneff: both, as was to be expected, very
unhappy and full of anxiety about the fate of France
and of Paris in particular.

I have had two very interesting letters from
Kingsley, the first very long and very eloquent. He
is *vehemently* German, and exults in the overthrow of
France. As to the French Revolution (the *fourth*
since I have been grown up), I rather wish than
hope or expect, that the experiment of a Republic

1870. may be really and permanently successful, and may show to other nations the example of a popular government compatible with *order* with respect for *law* and *authority*, and with the *security* of *property*. Without these, liberty is merely the license to do evil.

I hope the Germans will *not* retain Strasbourg, but only dismantle its fortifications, and those of other French towns within a certain distance of the frontier. If they mean to keep it, it is a strange proceeding to begin by destroying it, by so barbarous a measure as a bombardment. I am very sorry for Strasbourg,—sorry for the inhabitants and sorry for the destruction of the valuable Museum and Library, as well as for the damage to the Cathedral.

We go down (I hope) to Wells to-morrow. It has been a great pleasure to have Susan and Joanna's company hitherto.

With much love to all your family party,

Believe me ever,

Your affectionate brother,

Charles J. F. Bunbury.

JOURNAL.

Saturday, September 10th.

We took leave of dear Susan and Joanna.

Down to Wells, to visit our dear Arthur Herveys. They at present occupy the Deanery, as the Palace is new furnishing and fitting up.

They took us to see the palace, which is a grand and 1870.
venerable old building, in character somewhat intermediate, between a feudal castle and an ecclesiastical residence. The gardens beautiful ; we saw
them last year, and I think I noted in my journal at
that time the broad moat which incloses them, and
the noble remains of the old banqueting hall. Some
fine exotic trees : in particular, a superb Tulip tree,
an immense Acacia, a very large black Walnut, and
the largest tree I have seen of *the weeping large-leaved
Elm*.

The Tulip tree, 10ft. round at height of 3ft. ; the
Ailantus (which I at first took for a black Walnut),
8ft. 2in. ; a Salisburia (one of the finest and tallest
I have seen), about 4ft.

September 11th, Sunday.

At Wells.

Attended afternoon service in the beautiful Cathedral.

The Deanery is an old house. The room assigned
to me as a dressing room is known as King Henry
the Seventh's room, and according to tradition was
once occupied by him. The garden front of the
Deanery in particular, is venerable and fine, very
much like some fine old college at Oxford or Cambridge ; the garden also quite in character, with
velvet turf, noble trees, and old walls covered with
creepers.

Much delightful talk with the girls.

P

A very agreeable little expedition with the Herveys :—

Driving nearly to the village of Westbury (on the
road to Chedder) ; then walking up a pleasant green
dell among the hills, and up steep slopes of fine
close turf, sprinkled with old thorns and varied here
and there with limestone rocks, to a commanding
situation on the brow of the Mendips, from whence
the view was very extensive and beautiful. We
looked over the wide and fertile plain, beautifully
green and diversified by island-like hills, to the
Bristol Channel gleaming in the distance, the high
and remarkable promontory of Brean Down, the
bold islet of the steep Holme, the Welsh Hills seen
faintly on the horizon, the headlands near Minehead
and Porlock, the long high range of the Quantocks,
and more to the south, that of Blackdown. The
day was very favourable to picturesque effects :
variable, with frequent changing clouds and passing
storms, but also with occasional and partial gleams
of brightness. We descended by a pleasant lane,
in which the blackberries and sloes were in remarkable abundance and perfection, and rejoined
the carriages near the village of Easton, and returned to Wells by a different and very pretty road.
A dinner party. Canon Beadon was there, he is
ninety-three years of age.

———

Tuesday, September 13th.

Expedition with the Herveys to the famous
" Wookey Hole," and to Ebber Rocks. Wookey

Hole—opening in the face of a picturesque lime- 1870.
stone cliff, half mantled by wood, at the end of
a deep recess or "combe," in the Mendips;—a
small opening, not imposing like that of the Peak
cavern in Derbyshire. A paper mill supplied by the
stream which runs out of the cavern. Supplied
with farthing candles in the usual style, we all went
(a party of eleven besides the guide), to the end
of the cavern. Its characteristics much like those
of other caverns that I have seen: wet, rugged,
slippery, dark, dirty, and disagreeable; much
fatigue, some little danger, and nothing to make
it worth while, except (on this occasion) the company. The girls sang beautifully.

From this cavern we walked across a hill to
the Ebber Rocks, a sort of Chedder on a small
scale. A very difficult and fatiguing scramble down
almost a precipice into the ravine, which is richly
wooded in the lower part, and breaks into fine bold
limestone cliffs in the upper.

The entrance in the Wookey Hole (very small and
low), of the noted *Hyena den* was pointed out to
us by the guide; it is in the side of the ravine which
terminates in the cliff I have mentioned. The
locality is accurately described by Boyd Dawkins in
his paper in the 18th volume of the *Quarterly
Journal*, G. S.

Badgers still inhabit other crevices in these rocks.

A dinner party. Mr.* and Mrs. Church. Mr.
Church is a nephew of General Church.†

---

* Principal of the Theological College at Wells, and brother of the Dean of
St Paul's.
† He took an active part in Greece.—(F. J. B.)

A very pleasant trip to Glastonbury with the
Herveys; we ascended the Tor and drank tea very
merrily amidst the beautiful ruins of the Abbey.
Glastonbury Tor is a remarkable hill, commanding
the country for a long way round; the upper part
very steep, yet nowhere precipitous, but entirely
covered with close smooth green turf.   On the
narrow top stands a solitary tower—St. Michael's
Tower—very high, and seen a long way off, looking
very slender when seen from a distance, but really
of considerable dimensions and of very handsome
architecture.   The view is very fine: to N. and
N. E. the Mendip hills bound it at a short distance,
and their whole face is seen in great beauty, with
the pretty little city of Wells and its Cathedral in
full view, nestling at their feet.   In the other
directions the view is very extensive, and quite
panoramic, ranging over the wide alluvial plain,
diversified here and there by island knolls to the
Bristol Channel on one side, and on the others to
the various ranges of hills which bound it at greater
or less distance.   At our feet, as it seemed, the
Church and some of the houses of Glastonbury
peeped from amidst the wooded underfalls and
spurs of the Tor hill.   This hill with its underfalls
and spurs form in fact a real island (the "Island of
Avalon") in what is now fen and low alluvial plain,
but must, at a period geologically very recent,
have been water.   The city (or village) occupies
the western extremity of this island, extending
down quite to the edge of the marsh.   Brent

Knoll is a very conspicuous object in all the 1870.
distant views from the high grounds near Wells:—a
large, bold, round island hill, rising out of the low
alluvial flat near the shore of the Bristol Channel.

Thursday, September 15th.

We went through the Cathedral, carefully, with
Sarah Hervey and old Canon Beadon, who has
known it for 50 years, and who pointed out to us all
the details.

I noted last year the striking peculiarity of the
inverted arches at the junction of the nave and
transepts. and the droll sculptures in the capitals of
some of the pillars. The chapter-house is of
remarkable beauty : the manner in which the ribs
of the roof radiate from the central pillar, has been
justly compared to a palm tree. It stands high,
being over a crypt, so that one ascends to it (the
chapter house) by many steps from the body of the
Cathedral. In the cloisters (which are not as
strikingly beautiful as those of Salisbury), there is a
tablet to the memory of Mrs. Sheridan (the
beautiful Miss Linley) and her sister, Mrs. Tickell,
put up by their father (brother ?).

Canon Beadon, whom we first met at dinner at
the Deanery on the 12th, is a fine old gentleman of
93, with all his faculties in excellent preservation ;
an intelligent, well-informed man, who was well
acquainted with Wollaston. He is the nephew of a
former Bishop, and a relation of Sir Cecil Beadon.

A beautiful day. Excursion to Chedder Cliffs.
A very agreeable day in every respect.

Priddy, a small village on the bare limestone
plateau, has lead mines, which were worked by the
ancient Romans, and I am told it is found profit-
able now to work over again the refuse ores or slags
from which the lead was imperfectly extracted by
the Romans. The defile of Chedder is certainly
most remarkable, and the limestone cliffs and
pinnacles are wonderfully grand and noble. I
hardly think that I have seen finer rocks in any
part of Britain that I have visited. These towering rocks are here and there picturesquely mantled
with ivy or tufted with shrubs, and the slopes
between them are clothed with fine close green turf.
Pyrus Aria, the White Beam, (mostly in a shrubby
form) grows in abundance on the rocks, often
rooted in the crevices of the great cliffs ; I saw
Solidago (not abundantly) and the leaves of
Thalictrum minus : hardly any Ferns or Mosses.
The Chedder Pink, I believe, is now hardly to be
found except on the most break-neck crags.

Went to the afternoon service at the Cathedral.
Walked in the Palace gardens with the Herveys and
the Dean, and Mrs. Johnson.

We went with Sarah and Kate to the top of the

Tor hill, immediately adjacent to the city of Wells 1870.
on the E., and overlooking the whole of it. One
of the finest general views of the Cathedral is from
the ascent. The hill is composed of carboniferous
limestone, and there are large quarries in the south
face. We visited Mr. and Mrs. Church.

———— —

Tuesday, September 20th.

Mr. Church (who is head of the Theological
College here), showed to Lady Arthur and me
the Library of the Cathedral. A curious collection
of old books in a long low gallery over one side
of the cloisters ; there would not be room for a
great number of books, nor for a great number
of readers at once ; but under the conditions, it
is not a bad room for the purpose. There remain a
great quantity of iron chains by which the books
were formerly secured. The good Bishop Ken
bequeathed his library to this, and there are many
books with his name and his MS. notes in them.
Special curiosities shown to us by Mr. Church, are :
a fine copy of Aristotle and Theophrastus (printed
by Aldus, 1498), which belonged to Erasmus, and
has his signature and many of his MS. marginal
notes. A Commentary (by Joannes Damascenus),
which belonged to Cranmer, and has marginal notes
by him. Also some fine illuminated MSS. and
specimens of caligraphy of the time of Henry VII.
We spent the latter part of the afternoon delightfully, in a ramble with the Bishop and the
girls about a pretty little valley amidst the Mendips;

1870. first driving some three miles (of continuous ascent), along the Bristol road ; then leaving the carriages and rambling some way down the little valley and up the smooth green slopes beyond, where the ladies sat down and sketched. The weather, the scenery, the company, all delightful. We rejoined the carriages at the picturesque little village or hamlet of West Horrington ; here we found Asplenium Ceterach growing in abundance on the rough stone walls ; and descending thence by an almost precipitous lane, came into the Bath road at between two and three miles from Wells.

I have seen Asplenium Ceterach in several other places near Wells, on stone walls, not on rocks. Asplenium Ruta muraria, common, both on stone walls and limestone rocks, but I could not get any good specimens.

Wednesday, 21st September.

A most beautiful day. Joined Fanny and Sarah Hervey in the garden—they drew. I measured trees.

We went with the young Herveys (the four girls and their brothers), to the top of a hill called Dulcot, about two miles S.E. of Wells ; a very pleasant little excursion. The day was (like the two preceding) brilliantly fine and warm, but the distance was hazy, and the distinct view consequently not very extensive, though the nearer objects were beautifully seen. Dulcot is a remarkable hill, like a little mountain ; a separate ridge composed of highly

tilted strata of carboniferous limestone, very steep 1870.
on the south face, the strata dipping that way, and
still steeper (all but precipitous), on the north. We
ascended it at the western end, the one nearest
to Wells; this extremity of it is a very picturesque
object, which we had often admired from the Palace
gardens; the ascent is steep but not difficult, over
fine green turf, broken here and there by rocks.
Along the ridge the upturned and broken edges of
the strata stand out in bold crags, though on a small
scale; and the north face is thickly strewn with
fragments of rock.

In short, Dulcot is a serrated mountain in
miniature. It is peculiar in its character, so much
contrasted with the smooth rounded outlines of all
the neighbouring hills : and I do not see to what to
attribute the steep uptilting of the strata which
forms this single ridge.

————

Thursday, September 22nd.

Our pleasant visit to the Arthur Herveys came to
an end; we took leave, first of our dear Bishop and
Lady Arthur and Sarah, who went off for an engagement of long standing; afterwards of the rest of
the party, except sweet Kate, who went with us to
Kings Weston. A most enjoyable, delightful visit
it has been, everything conspiring to make it
agreeable. Our dear friends (though hard worked)
looking very well and full of all their old cordiality
and affectionate kindness to us. Altogether, I have
never known a family more entirely loveable and
admirable. The two elder girls are as charming as

1870. ever: the two younger are growing up to be like their sisters, and the two boys who were at home during our visit, (the two youngest, Arthur and Jemmy) promise to be worthy of such a home. It is, indeed, and has been ever since I have known them, a home of peace and love and goodness. We went, dear Kate accompanying us, to Bristol, by a line of railway passing through a very pretty country, by Chedder and across the extremity of the Mendips. At Bristol, Philip Miles's carriage met us, and conveyed us to Kings Weston, about 4 miles beyond Clifton.

September 29th. Thursday.

My Journal has fallen into arrear.

From the afternoon of the 22nd to the morning of the 27th, we remained with the Philip Mileses, at Kings Weston, and spent the time very agreeably. Philip and Pamela Miles were a very kind, attentive and agreeable host and hostess, and did everything in their power to help us to enjoy the beautiful country: the company of sweet Kate attributed much to our pleasure, and the weather was glorious. Kings Weston is a very fine place A large handsome, stately house, built of stone from a quarry on the property, with a grand entrance hall, very lofty, hung with portraits of the Tufton and Clifford families, to which the place formerly belonged. The house stands high, though screened on two sides by woods and higher ground: and to the W. and N. W., the park, which is of great extent, descends in beautiful sweeping curves and 1870. green undulations, varied by magnificent trees, to the low lands bordering the Severn. The view from the terrace before the principal front of the house is superb, extending far over the broad glittering Severn to the hills of Monmouthshire on the N., and westward across the mouth of the Avon to the picturesque headland of Portishead on the Bristol Channel. Although the weather was beautiful while we were here, there was a haze over the distance, which though perhaps it did not much diminish the beauty of the view, bounded its range ; if it had been clear, we were told, we should have seen the mountains of South Wales in one direction, and in the other the headlands far down the Channel.

The trees are magnificent, especially elms, beeches, Scotch and silver firs. There are some Scotch firs in the avenue opposite the S. front of the house, which are as fine as almost any I remember to have seen : and many other superb trees of the same kind in picturesque groups on the slopes of the park below the house. In the aforesaid avenue are two silver firs, (now evidently dying) of extraordinary height ; Philip Miles told me that one of them is 130 feet high, and I can well believe it : the other is not much less. Two beautiful beeches in the lower part of the park, measure each about 14 feet round, and are in full and undecayed vigour.

Kate and I had a long walk with Miles through his beautiful woods, clothing the ridge of hill east-

1870. ward of the house, and joining on to the woods of Blaize Castle. The luxuriant growth of evergreens is remarkable: above all, the arbuti are more numerous and larger and finer than I have seen any where else except at Killarney. Another day Pamela took us to Blaize Castle. The "Castle" is a modern sham, but it stands in a fine situation, high on the crest of a long wooded ridge, and the view from the top of the tower is very extensive and beautiful. In the side of this ridge of hill, towards the boundary of the Kings Weston and Blaize Castle properties, there is a very deep and striking ravine, richly wooded, but with fine walls of grey limestone rock breaking through the woods, much in the style of the banks of the Wye, near Chepstow and Tintern.

On the 26th we visited Sir W. Miles at Leigh Court, and were very courteously and hospitably entertained by Sir William and by his daughter Mrs. Tudway, who is a very pleasing person. Leigh Court is almost directly opposite to Kings Weston, with the Avon between them. The park and woods are very extensive, but, judging from what I saw, I should think not equal in beauty to Philip Miles's. The collection of pictures is justly famous; there is an extraordinary number of first-class works.

The 27th we left Kings Weston, and took dear Kate back to Wells. The Bishop and Lady Arthur were gone to London, but we had a glimpse of Sarah, and an affectionate leave-taking from both girls: and then started by railway for London,

where we arrived safe (thank God) a little before 1870.
seven.

Throughout these delightful seventeen days at
Wells and Kings Weston, we have been sur-
prisingly fortunate in weather : no rain, no cold
wind; and ever since the 17th it has been
beautiful and brilliant, many days, indeed, hot
enough at midday for the middle of summer, but
cold in the early morning and after sunset. No
rain since the 10th of this month, yet the pastures
in Somersetshire were brilliantly green.

Mrs. Byrne* is staying at our house, 48, Eaton
Place.

Wednesday, 28th September.

Weather still very fine.

Met Charles Lyell at the Athenæum, and had a
long talk with him. William Napier dined with us.

Thursday, September 29th.

Fanny went with Joanna to the Woking
Cemetery. I visited Sally, then to the Athenæum,
where I met Bentham and had a talk with him, also
Louis Mallet.

We had a family dinner party with the Charles
Lyells.

+ My Aunt, Mrs. Byrne. Mrs. Byrne left her house in Paris, on account of
the siege, and came to England. She remained at 48, Eaton Place all the
winter.—(F. J. B.)

Saturday, October 1st, 1870.

Julia Parkinson's wedding day, to George Betts.

Took leave of Susan and Joanna, who are to start to-morrow for Antwerp, on their return to Florence.

———

Friday, October 14th.

At Barton.

We went down to Folkestone on the 5th, and stayed there till the 10th, for the sake of sea air; during which time we had the pleasure of seeing much of William and Emily Napier and their children; returned to London on the 10th, and came down hither, to our dear home, yesterday, the 13th. Clement arrived from Geneva on the 8th and Edward from Switzerland on the 11th.

I have not, since the early part of September, put down anything in my Journal about public affairs.

The dreadful war goes on, and seems likely to acquire a more and more virulent and embittered character. Who, three months ago, would have dreamed of the possibility of such events as the siege of Paris by the Germans, and the Prussian head-quarters established at Versailles ?

The siege of Paris threatens to be one of the most terrible events in modern times. The destruction of such a city, with all the treasures of art, science, and literature which it contains, would be an irreparable loss to the whole civilized world; yet there seems little hope that it will be averted. The

Germans show little disposition to be merciful or 1870. forbearing in the use of their overpowering strength; they are determined to press their advantages to the utmost, and to insist on the cession of some of the richest provinces of France.

The French seem, on the other hand, to have worked themselves up to an almost ferocious spirit of resistance, and a determination to suffer anything rather than yield to such conditions. But I fear that the French have no army at all capable of opposing the Germans, and I cannot expect much good to their cause from a guerilla system, from an irregular desultory warfare, or from the isolated efforts of untrained partizans. Their *Francs-Tireurs*, and so forth, seem to me likely to have no effect, but to render the war more savage and sanguinary; and I shall be much surprised if Garibaldi* does them any good.

———

Friday, October 15th.

My Barton Rent Audit — very satisfactory; luncheon afterwards with the tenants and Mr. Percy Smith.

———

Sunday, October 16th.

A very wet stormy day. Read prayers with Fanny. Clement, in his way home from Geneva, visited poor unfortunate Strasbourg, and his account

See the correction of this opinion, December 29th.

1870. of its condition, agrees very well with that given in the *Times*. The devastation of the town by the bombardment is dreadful, and so must have been the sufferings of the people. One may rejoice that the Cathedral has escaped with but slight damage : but the Library, described with so much zest by Dibdin, and the Museum, so peculiarly rich in valuable specimens of fossil plants, are utterly destroyed.

The one redeeming point in this awful war, is the care taken of the wounded, and the grand efforts made for their relief, both by contributions in money and by personal exertions. In this respect there is a real advance since the Dark Ages.

Monday, October 17th.

Wrote to M. Jacquet to thank him for his care of Clement.

Edward arrived.

Wednesday, October 19th.

It is now some days since we heard of the sudden death of Edward Romilly ; and were much grieved at it. I had known him ever since I was a boy, though never *very* intimate with him. He was a wise and an excellently good man.

Monday, October 24th.

Wrote to Henry and sent an order to Drummond

for payment of young Henry's outfit. Dear Minnie 1870.
and Sarah Napier arrived, also John Hervey (the
Rev.).

A very fine Aurora Borealis was seen at night. It
first attracted our attention as we were sitting down
to dinner, about 8 p.m., and we watched it for some
time. Its colour was a beautiful red, between rose and
what is properly called aurora red, at times deepening
almost into crimson ; it was so glowing, that at first
sight I was inclined to fear that it was the reflection
from a great fire.* It was in the north-eastern
quarter of the sky, and had the appearance of a fine
red cloud, transparent, (for the stars were seen
quite distinctly through it), and not at all defined in
shape or outline: most brightly coloured towards
the centre, and gradually fading upwards and
towards the north into a fainter and diffused red
light, with some appearance also of white light
towards the lateral edges of the highly-coloured
part, and also below it. No very distinct corus-
cation visible as long as I looked.

Tuesday, October 25th.

Mr. and Mrs. Mills, Ladies Octavia and
Wilhelmina Legge, Leopold and Lady Mary Powys
and Minnie Powys arrived.

The Aurora conspicuous again this evening,
earlier than yesterday, indeed even before seven
o'clock: its colour almost as bright and red as

* In many places where it was seen, it was taken for the indication of a
fire, and in London, it is said, the fire engines of some parishes were ordered
out. It was also thought by some to be the burning of Paris.—(F. J. B.)

1870. yesterday.   It was seen in the W. as well as in the E., and seemed to extend in a grand arch all across the sky.

_____

Mrs. Rickards came to luncheon.—I had not seen her for a long time.   John Kinloch and some other young men arrived—also Admiral Spencer.   Lady Cullum and General Ellice dined with us.

_____

Walked round the arboretum with Mr. Mills.

_____

Rested.   Fanny met with an accident at Rushbrook (a fall), which prevented her from going to the Bury Ball: but all the rest except Mr. and Mrs. Mills and myself went thither.

News of the surrender of Metz.

_____

George Napier arrived.   I am very glad to learn from the _Gardeners' Chronicle_, that Professor Schimper escaped from Strasbourg just before the siege.   It is not stated, however, _where_ he now is, or is supposed to be.   I fear that his fine collections must certainly all have been destroyed.

_____

LETTER.

Barton, Bury St. Edmund's,
October 29th, 1870.

My dear Katharine,

Our dance on Thursday night went off very well 1870. indeed, and was very pleasant. There were nearly 100 people, including those staying in the house: many pretty girls and pleasant people; the dancing was kept up with great spirit till about three in the morning, ending in a long cotillon. Sarah Napier (who looks as lovely as ever), was "Mistress of the Revels." I danced the first quadrille with Lady Mary Powys, and Sir Roger de Coverley with Lady Cullum;—frisky for me, was it not? I missed the dear Herveys, but we had (and still have) a charming party of girls in the house. Everybody looked pleased and happy. The party staying in the house were, besides ourselves and Edward and Clement: —Minnie and Sarah, our dear Mr. and Mrs. Mills, Lady Octavia and Lady Wilhelmina Legge, Mr. and Lady Mary Powys, Miss Powys, Admiral Spencer, Miss Praed and her brother, John Hervey, John Kinloch, Charles Young, Mr. Willoughby Burrell, and two or three more young men. Mr. and Mrs. Gilstrap brought a large party, among whom was M. Gudin, the famous French sea-coast painter, several of whose pictures I had admired in the different exhibitions and in the Luxembourg. He married a Scotch lady, and speaks very good

1870. English; he told me that his son is serving in the National Guard in Paris.

Did you see the Aurora Borealis on Monday and Tuesday ? It was very beautiful; though I did not see any distinct coruscations, but a wide-spread glow of beautiful rose colour, deepening at times almost into crimson. When we first caught sight of it (just as we were sitting down to dinner on Monday), we thought for a moment that it was the reflection from a very great fire. It is a great comfort that William Napier's children are going on so well.

(October 30th.) I hope you and your children are well also, and I hope that Rosamond's pets are flourishing. All Fanny's are very much so, and her little parrakeet is a most amusing creature. I have just got Lubbock's "Prehistoric Times," the new edition, and propose to begin soon to read it through.

Ever your very affectionate brother,

CHARLES J. F. BUNBURY.

JOURNAL.

November 1st, Tuesday,

Dr. and Mrs. Phelps (of Sydney College Cambridge) arrived yesterday.

There was a great fire on Mr. Loft's farm.

Wednesday, 2nd November.

We went to visit the scene of the fire, and I wrote to Major Heigham, the Chief Constable.

Mr. and Mrs. Walrond arrived.

A visit from Major Heigham on business.

———

The Miss Richardsons, Mr. Maurice and George Darwin arrived.

———

We went to afternoon Church and heard an admirable sermon from Mr. Maurice. Read in the evening to the household Kingsley's beautiful sermon on Pardon and Peace.

———

Had a delightful walk with Mrs. Walrond. Mr., Mrs. and Miss Fuller Maitland arrived. The Milbanks and Lady Cullum dined with us.—Much laughing.

———

Lord Charles Fitzroy and his son came to luncheon. The Hortons dined with us.

———

Minnie Powys went away. Mrs. Hardcastle, Sir James and Lady Scarlett and Mr. Peter Scarlett came to luncheon.

———

Mrs. Walrond, before breakfast, sang me a beautiful Cornish song.

1870.   The Miss Richardsons and the Fuller Maitlands
went away in the morning.   The Walronds, Edward
and Clement in the afternoon.   The Bishop's conference at Bury.

November 11th.

It has been a most agreeable time.   Mrs.
Walrond is even more charming than I thought her
in 1868.   Mr. Walrond is very agreeable, gentle,
quiet, cheerful and very well informed.   They seem
a most devoted couple.

The Miss Richardsons, as I have always found
them, extremely pleasant.

LETTER.

Barton,
November 11th, '70.

My Dear Katharine,

Our pleasant party is now all broke up and
dispersed, none remaining here but Minnie and
Sarah and George Napier : and the last two will
leave us to-morrow.   The last ten days have been
very pleasant, the only drawback being dear
Minnie's illness, which for several days past has
entirely confined her to her room : but I trust she
will be quite well before long.

My last letter to you brought our *history* down to
the 29th of last month,—the day after that on
which Fanny had her fall.   Happily she has
suffered no harm from it.

Colonel and Mrs. Lynedoch Gardiner arrived on 1870.
the 1st: Mr. and Mrs. Walrond on the 2nd: the
Miss Richardsons, Mr. Maurice and George Darwin
on the 5th, on which day the Millses and the young
Lady Legges went away. The Walronds, Edward,
and all the rest except the three Napiers left us
yesterday. Mrs. Walrond (of whom I daresay you
may have heard Lady Bell speak) was in high
beauty and spirits, and I admire her more than
ever. I have seldom met a more fascinating person.
Her husband is also very agreeable. The
Gardiners are very· pleasant, and I found a *sister
botanist*, and a very zealous and intelligent one, in
Mrs. Gardiner: she confines herself, indeed, to
British plants, but understands them very well. I
showed her part of my herbarium. Mr. Maurice
and the Miss Richardsons you know well, so there
is no need that I should praise them. We were
both of us much pleased with George Darwin.

On the 1st, and again on the 4th, the four girls
with John Hervey and Clement, acted very prettily
a little play of their own writing — a comic
cinderella : excessively pretty and amusing.

(November 12th.) Our darling Sarah has just
left us, with her Uncle George, and we are left to
ourselves, to our books and accounts and home
occupations. I must honestly confess, that after so
long a spell of such agreeable dissipation, I feel just
now rather *sick* and *spoony* and unsettled :—much
as, when I was young, I used to feel on the morning
after an agreeable ball. But I shall very soon
settle into my regular course of reading and

1870. occupation, and we have plenty to do, both indoors
and out, especially with reference to our schools,
which the new Act makes it necessary to look
closely to.   Also, a good many trees in the
arboretum must be sacrificed for the benefit of the
rest.

Your letter of the 31st gave me great pleasure,
and I am very glad to hear that you have got some
valuable additions to your collection of ferns.   I am
reading Sir John Lubbock's "Prehistoric Times,"
which I find remarkably clear and well written, and
less dogmatical than I had expected; also the
"Life of Lord Palmerston" by Sir Henry
Bulwer.

I am very *very* sorry for the Tourgueneffs: for all
who have friends or property in Paris, the prospect
is most painful and deplorable: indeed, it is painful
enough to all who have any feelings of humanity.
I fear the siege of Paris will figure in history as a
counterpart to the siege of Rome by Alaric.   We
may well be thankful that Mrs. Byrne escaped
from it in time.

Believe me ever,
Your affectionate brother,
CHARLES J. F. BUNBURY.

JOURNAL.

On Saturday, the 12th November, dear Sarah and <inline type="margin-note">1870.</inline> her Uncle George went away—a sorrowful parting. I drove out with Fanny to Euston, and called on the Duke of Grafton and the Phippses.

———

Monday, 14th November.

I was allowed to see dear Minnie.

Resumed the catalogue of my herbarium—Compositæ.

As well as I can make out, the great military lesson of this war seems to be the omnipotence of artillery ;—that to be superior in artillery is to be superior in war. Of course I do not mean merely superiority in number of guns or in construction of guns ; but superiority also in judgment, in placing them, as well as in skill in working them.

Under these conditions, the power of modern artillery seems to be irresistible ; witness the battle of Sedan. And from hence seems to follow a terrible corollary,—that the possession of such an artillery gives an irresistible power to a government—that there is henceforth no hope either for an insurgent population against a government, or for a weaker nation against a stronger one.

The great political lesson to be drawn from the collapse of French power is, I think, this :—that a corrupt, deceitful, immoral system of government, —a system of deceiving and misleading the nation,

1870. is sure, in the long run, to weaken and undermine every department of the State, and to end in ruining the ruler who has trusted to it.

———

Tuesday, 15th November.

We dined with the Benyons at Culford. Met the Maharajah and Maharanee, Mr. Cox and Lady Wood, Sir Edward Gage, the Hortons, Bains, and others.

——— —

Wednesday, 16th November.

Mrs. Ellice and her two daughters arrived. Lady Cullum, the Suttons, Mr. Percy Smith, Mr. Murray, and Mr. Fitten, dined with us.

———

Friday, 18th November.

Minnie came down to dine with us, the first time for a fortnight. I had a very pleasant talk with her.

———

Monday, 21st November.

Wrote to Clement, sending him an order for money to pay his conveyances.

Resumed my work at Life of Sir John Hawkwood.

Discussion with Fanny and Scott about schools at Mildenhall.

———

Wednesday, 23rd November.

A pleasant walk with dear Minnie about the arboretum.

A visit from the Duke of Grafton.
Sir Henry and Lady Tombs arrived.

Thursday, 24th November.

Sir Henry and Lady Tombs went away after breakfast, and after seeing the cottage.

Dear Minnie Napier left us this afternoon, on her return home; we were very sorry to part with her. We both love her as a sister.

Sir Henry Tombs said to me yesterday, that he cannot make out that the Prussians in the battles of August and September had a greater number of guns, in *proportion to their number of men*, than the French. From all he can learn, the proportional number of guns was about equal on the two sides, and it is not clear that the Prussians used guns of a superior quality; but they showed very superior skill in placing them, and especially in *massing* them —not scattering them along the line,—so as to concentrate an overwhelming fire on the important points of the French positions. This, I think, was an object to which the first Napoleon was always very attentive.

Sir Henry said that the Prussian guns were breech-loaders, the French muzzle-loaders; but he does not think this had much share in the victory of the former. The French chassepot rifle, he says, is undoubtedly superior to the needle gun, but the better discipline, the superior coolness and steadiness of the Germans, more than made up for the inferiority of their weapons.

1870.   Three days ago I received from Dr. Thomson the sad news of the death of my nephew, Henry Fox Bunbury, at Meean Meer, near Lahore, in the Punjaub. He died of remittent fever in October, aged 31. It is a very melancholy event. He has been in India continuously for upwards of eleven years; and for several years before he went thither, I had seen him but very seldom; so that I really had very little personal acquaintance with him, since the time when he was a mere boy.

I read to Fanny this evening some part of Milman's essay on Erasmus.

Monday, 28th November.

Received a long letter from Henry on the death of our nephew Harry.

Wednesday, 30th November.

Visits from Mr. Lott and Mr. Isaacson; with whom and with Scott we had a long consultation about Mildenhall schools.*

Sunday, 4th December.

We went to morning Church and received the Communion. There was an offertory on behalf of the French peasantry.

---

* For the last three or four days Sir Charles had been making notes in reference to the Mildenhall Schools.

Monday, 5th December,   1870.

Lady Rayleigh and Miss Strutt arrived (for one night).   The Abrahams and Mr. Murray dined with us.

─────

LETTER.

Barton,
December 7th, 1870.

My dear Katharine,

Many thanks for your interesting letter of the 29th November. We went over to Mildenhall on the 1st, and stayed there two days ; on the 2nd, I presided at a public meeting about the Education Act. My address was well received, and the meeting showed a unanimity which surprised me ; indeed there was very little discussion, and all the resolutions proposed by Mr. Isaacson were carried unanimously. We had already come to the conclusion that there was little chance of making up the deficiencies by voluntary subscription, and that it would be best to apply to the Education Department to give us a Board which should build and maintain the necessary schools by rates, without interfering with our existing schools. So it was decided.

We found dear good old Mrs. Marr flourishing, and seeming happy and comfortable ; and the house in very nice order, but excessively cold. Fanny took her little paroquet over there with her, but not to the meeting !

Since we returned, I have been looking over the

1870. herbarium (of British plants) of our old friend
Mr. Rickards, which Mrs. Rickards has lent me.
It is rich in number of species, very well arranged,
and in excellent preservation, and contains many
very rare plants; indeed several of which I had not
before seen, British specimens; such as Lysimachia
thyrsiflora, Senecio paludosus, Asperugo procum-
bens, Epipactus ensifolia, and others.

[Since I wrote this yesterday, Mrs. Rickards has
most kindly and generously made me a present
of the whole collection].

December 8th. Mr. Murray, the publisher, has
sent Fanny the newly-published handbook of the
Eastern Counties, to which she had contributed
a great deal of information about Barton and this
neighbourhood; it is very well done as far as I have
looked. I have been reading to her in the evenings,
Dean Millman's essay on Erasmus, and we are
much pleased with it. I am delighted to see justice
done to Erasmus, who is a man I have a great admiration and sympathy for, but even Milman's essay
has, for me, a litte too much of an apologetic tone.

I am just finishing Lubbock's "Prehistoric
Times" (the new edition), the latter chapters, on
modern savages, are very curious and entertaining.
How inexhaustibly various and strange the forms of
human folly and absurdity are! The manners and
customs of the Feejeeans seem particularly objectionable,—at least they appear so to us, who are in
the decline of life, and yet do not see the necessity
of our being put out of the world. The missionaries
have done so much in shewing the Feejeans and

other savages the error of their ways : I wish they 1870. would try their christianizing skill upon European kings and ministers. There seems, indeed, now to be good hope that the Russian storm may blow over. It would be too horrible if the whole of Europe were involved in war, converted into one vast howling den of wild beasts. Little Arthur Mac-Murdo might well ask (as he did ask Fanny), whether " people love their enemies when they kill them in battle ? '"—when they tear them to pieces with shot and shell ? I think the prospects of civilization in Europe are darker now than they have been since 1815. The enormous successes of Prussia in this war will lead the Powers to devote their energies more absolutely than ever to military objects and to keeping up monstrous armies. The great military lesson of this war is, the supremacy of artillery :—the government which is superior in artillery will be superior in war : and thence follows a formidable corollary, that there is henceforth no hope, either for an insurgent people against a government, or for a smaller nation against a greater.

Love to your husband, Rosamond and your boys, and to Charles and Mary,

Ever your affectionate brother,

CHARLES J. F. BUNBURY.

═════

JOURNAL.

Saturday, 10th December.

Mr. Brooke of Ufford came to see our pictures,

1870. library, &c., and spent much of the afternoon with us. We had a visit from Lady Hoste.

Tuesday, 13th December.

Resumed the Hawkwood MSS.

LETTER.

Barton,
December 18th, 1870.

My Dear Edward,

I suppose you will soon be departing for your usual Christmas resort, so I write a line to wish you a merry Christmas and a happy New Year, and to inform you that we are alive and well. We are leading very quiet and regular lives, with very little variation, only last Thursday we dined at Hengrave, as it happens, the first time I have dined in that house since the days when old John Powell was living there. And the next day a large party from Hengrave, consisting of Lady Gage, the Milbanks, Lady Augusta Cadogan, your friend, Mrs. Ford, Sir Henry and Lady Ibbetson came here to luncheon and to see the pictures, our grandfather's drawings, &c. : Lady Cullum joined them here. It was a pleasant party.

How many thousands—hundreds of thousands there are—to whom this Christmas and New Year will be a season not of merriment and happiness, but of suffering and misery and horror. And who can tell what *our* condition will be by this time next year? I was *very* sorry the French *sorties* were not successful. But though the prospect does not look very bright for them, it does not seem entirely hopeless that they may succeed in wearing out the Prussians, and at any rate they have amply redeemed their character for courage. The way in which Paris has held out, and the persevering spirit with which they have fought the Germans again and again, renewing the struggle continually, is, I think, more really honourable to the national character than their victories under the first Napoleon. When the German armies first arrived before Paris, I certainly did think that the case was hopeless for the French, that it was useless for them to try to hold out any longer, and could only lead to mere waste of lives: but I was completely mistaken. Whether they will be successful in the end, I cannot tell, but they certainly have well earned the respect of the world by their resistance. It is as fine as that of the Spaniards in 1808 and 9. If you hear any news which is not in the papers, I hope you will let us know.

The season seems to be an unhealthy one. Image says there is much sickness about. I hope you will take care of yourself. With love from Fanny,

Believe me,

Ever your affectionate brother,

CHARLES J. F. BUNBURY.

JOURNAL.

1870. The great eclipse of the sun. Although there were heavy falls of the snow, both in the morning and the afternoon, the middle of the day was fine and clear, and we saw the eclipse well. Though only a partial one as seen by us here, it was a striking and interesting sight. It was particularly well seen about 12 o'clock, and till about a quarter after, when light thin clouds passing over it, took off its dazzling effect, and at the same time left its outline perfectly distinct, so that we could see it quite well without the aid of smoked glass. Its apparent form then was completely that of a crescent, or perhaps more correctly that of the waning moon: its convexity being turned towards the right. About 12.30 the "horns" of the crescent appeared to point almost directly downwards. I did not observe any remarkable darkness, but the clear part of the sky appeared of a peculiar tint of blue.

I received Lyell's new book, "The Students' Elements of Geology," and looked through a part of it.

A thorough old-fashioned Christmas. These last two days have been like the Christmas of 1860 :— bright, clear, sunny days, with strong sunshine, no wind, the ground covered, (though not very deeply)

with snow, and intense frost.   If the thermometers 1870.
in our garden are to be trusted, the cold of the last
two nights (the 23rd and 24th) has been 26 deg. of
Fahrenheit, below freezing point, or 6 deg. above
zero of Fahrenheit.   To-day there is not a cloud in
the sky, and the surface of the snow glitters with
strange brilliancy in the sunshine.

Monday, 26th December.

Cecil arrived : also Mademoiselle Pierredon (to be
governess to Arthur during the holidays).

Tuesday, 27th December.

More snow and clear frost ; the servants' dance
and supper.

Thursday, 29th December.

A very fine day, the frost continuing, and ground
covered with snow.

Went out for the first time, since the 20th, round
the arboretum and garden.   In the early part of
this present month, our dear old friend, Mrs.
Rickards, most kindly and generously made me a
present of her whole herbarium, collected in the
course of many years (beginning from 1834) by
herself, her husband and daughter.   It is a rich and
valuable collection of British plants, excellently well
arranged, carefully named and in excellent preservation.   It includes a great many rare species,
especially several old specimens given by Mr. Hasted

1870. from his herbarium, which are now scarcely to be
found, such as Senecio paludosus, Senecio
(Cineraria) palustris, Melampyrum arvense, from
Costessey, near Norwich), Vicia lutea (from
Oxford), Hypochœris maculata.

The year is near to its close, and again it is my
wish, as well as my duty, to offer my most earnest
and humble thanks to the Almighty Giver of all
good, for the many blessings He has permitted me
to enjoy. Throughout this year, as well as the
preceding ones, I have been permitted to retain all
these inestimable blessings for which I have often
before expressed my thankfulness; above all, such a
wife, and such friends, that no man ever had
better. My health has been, without interruption,
good, and I have especial cause to be thankful for
my escape (on the 10th of April) from a very serious
danger.* My worldly circumstances have been as
good as possible, and I am thankful that I continue
on terms of perfect peace and good will with all my
dependents, tenants and neighbours. But in one
respect this year has been sadly distinguished from
the previous ones for a long way back. A dark and
awful shadow has been thrown over us all by the
sorrows of others—by the vast mass of accumulated
sorrow and misery in other countries. It has been
impossible not to feel one's mind oppressed, and
one's heart afflicted, and one's thoughts disturbed,
by the horrible war which has for these five months
devastated France. It has seemed at times almost
wrong to be cheerful and happy while misery on so

---

* He escaped being drowned in a pond in which he was searching for plants.

gigantic a scale was prevailing among our neigh- 1870.
bours. It is frightful to think of: indeed the
imagination cannot realize the amount of human
suffering caused by this wicked war. What an
illustration of "Peace on earth and good will
towards men."

The prospect of the coming year (as to public
affairs I mean) is very dark and dreary. I do not
think the hopes of civilization and Christianity in
Europe have appeared so over-clouded, so dim, at
any time within my memory. It almost makes one
despond. It looks as if force and violence were
destined to be for a long time the only rulers of
Europe. The prospect for our own country is very
threatening, and I certainly cannot draw any comfort from the wisdom of our present Government.

––––––––

December 30th

My nephew Cecil, who is lately come from Gibraltar, tells me that the apes inhabiting the rock are
now reduced to one family consisting of only nine or
ten individuals. The authorities endeavour to
preserve them, but the goat-herds, who ramble over
the most remote and wildest part of the rock, cannot
be prevented from taking the young ones. He tells
me also that vultures have become scarce on the
rock. He has seen abundance of that beautiful
bird, the bee-eater, in the neighbourhood, especially
in the famous Cork wood. Bustards inhabit the
neighbouring parts of Spain.

––––––––

The year is closing with great severity of weather.
Fortunately there is little or no wind, but the frost
is intense ; in the night before last, the thermometer
in our garden went down to 12 deg. Fahrenheit, and
last night to 10 deg.   The snow is not deep, but
continuous with a crisp and glassy surface.

This day has been very agreeably signalized by
my receiving a charming letter from Sarah Hervey.
Her letters are always admirable.

There is no part of this year, to which I look
back with keener pleasure, than the time we spent
with the dear Herveys at Wells.

The period from the 24th October to the 12th of
November, while this house was full of cheerful
and agreeable company, was likewise exceedingly
pleasant.

We have lost, in the course of this year, a nephew
Henry Bunbury ; a cousin by marriage and an
intimate friend, Frederick Freeman ; and an old
friend and very estimable man, Edward Romilly.
Of our acquaintance not so many have departed as
last year.

And so ends the year 1870 ; a year which I
personally shall ever remember with pleasure and
gratitude ; but an awful year of sorrow and suffering
to millions of our fellow creatures.   God grant that
the coming year may be better and more peaceful.

1871.

LETTER.

My Dear Leonora,

I do not know whether it is strictly 1871. orthodox to wish a happy new year *after* New Year's day; but nevertheless you will allow me to do so, for I do sincerely wish that this year may be a happy one to you and your husband, and your dear girls, and to all whom you love.

This opening year finds us both (I am thankful to say) quite well, in spite of the severity of the weather; for we have a regular old-fashioned winter, such as one reads of in Thomson's "Seasons," and such as I can remember in my childhood; steady, hard frost, more intense than we have had since the winter of 1860-61, with snow not deep indeed, but uniformly spread, and often bright sunshine. I fear it will tell upon the evergreens. Fanny has to take much care to preserve her pet birds from suffering from the cold, and as yet none of them appear to have taken any harm. Her little Australian paroquet Zoé, is flourishing, and is the tamest, most impudent, most caressing, most amusing little creature you can imagine. For myself, I do not enjoy the cold at all, but am thankful to have a warm house, and to be able to sit by the fire and read and write.

I have lately received Lyell's new book, " The

1871. Students' Elements of Geology," which, no doubt, you will see; it is a neat, compact volume, containing a marvellous quantity of knowledge crammed within a moderate space. It is sure to be very useful. I have lately finished reading Mackintosh's " Dissertation on Ethical Philosophy," which indeed I had read twice before, at intervals of many years; it is an interesting and beautiful book. I have now begun to read a little of Shaftesbury's "Characteristicks," a noted old book; the copy which I have, belonged to your Uncle.

Two of the principal books I have read this winter, are the "Life of Lord Palmerston," by Sir H. Bulwer; and Sir John Lubbock's "Pre-historic Times" (the new edition). Both well written and very pleasant reading.

The first volume of "Faraday's Life," also is very interesting, and shows in a very clear light, his noble and beautiful character; but the second is quite beyond my comprehension. In my specially favourite department of natural history travels (like those of Bates and Wallace) I have heard of nothing new for a long time.

I cannot say much about politics, though I must say that I was much struck by the candour and generosity of your remark in your last letter; but that war is a subject hateful to me; I can hardly bear to read the newspaper. There appears no present hope of peace, and the longer the war goes on, the more envenomed and fiendish it becomes. It almost makes me despair of the prospects of civilization and Christianity in Europe.

What a Christmas this has been.—What an 1871.
illustration of "Peace on earth and good will
towards man." In this terribly severe weather, the
suffering of the armies must be very severe, but I
feel most for the poor unfortunate peasants, who
have been burnt out of house and home.

With love to Annie and Dora,

Believe me ever your affectionate brother,

CHARLES J. F. BUNBURY.

JOURNAL.

January 5th.

Our dinner party—the Abrahams, Lady Cullum,
Hortons, Mr. Beckford Bevan and his daughter.

January 6th.

The Abrahams stayed till the afternoon, and
were very pleasant. Mr. Abraham tells me that
several specimens of the mountain finch have been
seen in the last few days in his parish of Risby.

Clement saw a fine hawfinch the other day in a
hedge between this and Livermere.

This severe winter may naturally be expected to
send us some uncommon birds.

January 10th.

Our dinner party—the Hardcastles, Lady Hoste
and Mr. Green, Lady Cullum, Patrick Blake, the
James Blakes, Captain and Mrs. Browning, Mrs.
Coddington, General Ellice.

January 11th.

Mr. and Mrs. and Miss Praed came, and Mr. Abraham. Lady Cullum also dined with us.— Afterwards Fanny and all the party except Herbert and I, went to Mrs. Wilson's ball.

————

January 13th.

Here is an old epigram, which Lady Cullum told me the other evening. It was on Dr. Goodenough (the Bishop of Carlisle, and friend of Sir James Smith) preaching at Court:

> " 'Tis well enough
> That Goodenough
> Before the Court should preach,
> For sure enough
> They're bad enough
> He undertakes to teach! "

————

January 14th.

I was sorry to see in the newspapers yesterday, the death of Henry Alford, the Dean of Canterbury. We never were very intimate, but I knew him pretty well at Trinity, and have met him occasionally since, and he was very attentive and pleasant to Fanny and me, when we were at Canterbury the year before last. He was a good, a learned, and a useful man.

He was a year my senior in standing at Cambridge, though a year my junior in age. The regret for his death seems to be very general and unanimous.

————

January 16th.

Read the story of " Pyramus and Thisbe," in Ovid's " Metamorph."

January 18th.

Mr. and Miss Loch arrived.

January 19th.

Walked round the grounds with Mr. George Loch who is very pleasant.   Fanny and all the party except Herbert and myself, went to the Bury ball.

January 20th.

The Lochs went away.

I read the stories of Salmacis and of the Miny-eides, in Ovid's Metamorph. lib. 4.

The Hortons and Thornhills, with a large party, came in the evening, and the young people had a dance.

January 21st.

Miss Byng and Miss Praed, Captain Eyre and John Hervey went away.   Lady Mary Egerton and Miss Egerton arrived; and late in the evening our dear Arthur Herveys arrived quite well, after their long journey from Wells.

Sunday, January 22nd.

We went to morning church with the Arthur Herveys and a large party.

Read Kingsley's excellent Sermon on God's World.

Lord and Lady Augustus Hervey dined with us. —Lady Augustus's singing wonderful.

——— ——

January 25th,

Lady Cullum, Lady Gage, Patrick Blake, Mr. Borton, the Archdeacon and Mrs. Chapman, Mr. Browne, and the Wilsons, dined with us.

——— ——

January 28th.

We have passed a truly delightful week, in the enjoyment of the society of our dear friends the Arthur Herveys. They came to us,—the Bishop and Lady Arthur, Sarah, and Kate, and Constantine (the 4th son),—this day week, late in the evening, having travelled through from Wells ; and they left us this afternoon. The Bishop seemed particularly well and happy.

We have had some other agreeable company. Our dear Minnie and Sarah came on the 17th and stayed till the 23rd. Sir Francis and Lady Boileau from the 18th till the 23rd; Lady Mary Egerton and her eldest daughter from the 21st to the 26th. Lady Boileau is remarkably pretty and very pleasing.

LETTER.

Barton,
January 29th, '71

My Dear Katharine,

We have had a delightful week in spite of 1871. the severe weather, enjoying to the full the society of our dear Arthur Herveys, who came on the 21st and went away yesterday: too short a time, but very delightful. The Bishop and Lady Arthur are looking particularly well, and nothing could be more cordial and genial: and the girls as charming as ever. We had also, during part of the time, some other very pleasant people with us, particularly Lady Mary Egerton and her daughter and Mr. Milman. Now we are alone for a little while, *i.e.*, there is nobody here except Herbert and little Arthur, but we are looking forward to seeing Charles and Mary this week, unless (which I heartily hope may not be the case), the severe weather makes them afraid of coming. It is certainly a very hard winter. I hope neither you nor yours have suffered any harm from it. I am very thankful that both Fanny and I have continued as yet quite well.

I was very glad to hear that you had added to your collection two such rare Ferns as Cystopteris Sudetica and Asplenium Seelosii: the latter I know by the figure in the "Second Century of Ferns," the other not at all. I suppose you must now have nearly all the European ferns. There were none remarkable in Mrs. Rickard's collection, except Adiantum Capillus Veneris from Ilfracombe.

1871.   I amused myself not long ago by making a list of
all the *wild* plants which grow in this park, including the groves and hedges which immediatly
surround it : they amount to 218 flowering species,
of which 35 are grasses. This number includes a
few which are probably naturalized, but thoroughly
established, such as two crocuses, a tulip and two or
three others.

I have been making much use of Hooker's
" Student's Flora," and the more I use it the more
I approve of it. In fact, I think it is *the* British
Flora for everyday use, speaking of *non*-illustrated
works on the subject. With this and Charles
Lyell's new Elements, a tourist will be capitally
equipped.

I have heard of nothing for a long time past in
my favourite line of reading, namely " Naturalists
Travels." I daresay Leonard will some day con-
tribute to that branch of literature, but probably it
will not be in my time.

I was a good deal surprised that Huxley, whom I
had always supposed to be rather an overworked
man, should seek the additional labour of a seat on
the School Board.

With love to your husband and children,

Believe me ever,

Your affectionate brother,

Charles J. F. Bunbury.

JOURNAL.

The capitulation of Paris, of which the news has 1871. just come, is a mighty event, and affects one with a variety of emotions.   On the whole, I am glad of it, under all the circumstances, as it will save the beautiful city from destruction, with all its treasures of art and science, and save its people from further misery and slaughter.   Indeed, one must rejoice at it very much, one must feel it quite a relief when one thinks of all the suffering and misery which the siege has already caused.   And as there seemed to remain no hope of driving away the Prussians, it was very right to put an end to those miseries by surrender.   But yet I feel for the humiliation of the French, and wish that they *could* have driven back the enemy.

Will . this capitulation and armistice lead to peace ?   It is much to be wished, for success on the part of France appears to me hopeless, and to prolong the hopeless struggle would only be to prolong and aggravate all the horrors of this most horrible war.   But I greatly fear the contrary.   I fear that Gambetta (whom I believe to be as thorough a fanatic as Robespierre or St. Just), will do his utmost to make the people of some part of France continue the fight, and that the passions and fanaticism of the French have been so thoroughly roused, that they will be only too readily influenced by him.

1871. The best hope is that the armistice and the election of a representative Assembly may allow the warlike passions of the French to cool.

February 1st.

Finished Hume on the "Passions," and began to read his larger essay "On the Principles of Morals."

February 2nd.

We dined with Mr. Beckford Bevan.

February 4th.

My 62nd birthday :—thanks be to God.

February 11th.

Very pleasant company in the house this week: the Charles Lyells, Lord Charles and Lady Harriet Hervey, Lady Head and her daughter, Mr. Twistleton,* Mr. Strutt, Alfred Newton, Mr. Pryor, Mr. Bowyer. I like Lord Charles Hervey particularly : he has a strong likeness to his brother Lord Arthur, with much of the same charm of manner, and (as well as my knowledge of him yet enables me to judge), much of the same sweet, refined noble nature. He returned less than a year ago from a voyage to South America and the

---

* Mr. Twistleton's story about my uncle Henry Fox, at Washington, invited to a wedding, in danger of being too late, could not get a carriage, saw an empty *hearse* passing, bargained with the driver and got in; amazement of the wedding guests when they saw the hearse stop at the door, and the British Minister emerge from it.

Sandwich Islands, and he has in previous years 1871.
visited, for his health, many other distant countries;
thus he has seen a great deal, and not only *seen*, but
observed, and observed very well : and he talks
readily and well of what he has seen. His conversation is very good. Though not profoundly
scientific, he has a great love for natural history,
especially botany, and has observed very well.
His experiences of earthquakes in his recent travels
were copious and interesting ; the first he told us,
was at Valparaiso, where he had sat up late at night
reading some papers on business; he had just
finished when he felt a shock which seemed as if a
huge giant had jumped with all his might on the
floor above ; it shook the whole house, and then the
vibrations seemed to be prolonged with diminishing
intensity into the distance. When he was in
Hawaii (Owhyhee) he felt earthquake shocks almost
every day, and often more than once in a day. But
the most violent he experienced was at San Francisco, in California ; there he was staying in a large
hotel, a solid well constructed building, and he saw
the walls of his room *rocking* like the bulkheads of a
ship's cabin in a heavy sea, and looking out he saw
the opposite walls of the court yard swaying in the
same manner ; it seemed as if the whole must come
down in one mass of ruin.

Lord Charles spoke strongly of the prodigious
abundance and luxuriance and beauty of the ferns
on the mountains of Hawaii. In the upper zone of
vegetation on those mountains, he said, they seemed
to predominate over all other plants. He observed

1871. that ferns were the first plants which appeared on the recent lava, springing up in the still smoking crevices.

He ascended the two very high mountains of that Island, Mouna Loa and Mouna Kea, and saw the famous "lake of fire," Kilanea.   Lord Charles said that in the ascent of these great mountains in Hawaii, he suffered much from that loss of muscular power and difficulty of breathing occasioned by the rarity of the air at great heights, and often mentioned by travellers, especially on the Andes.

In California he visited the Yosemite valley, and the "Mammoth" trees·   By the way, I learned from him the right pronunciation of the name of that valley, which had often puzzled me.   It is a *quadrisyllable*, with the accent on the second: Yo-sĕ-mĭ-te.   He confirmed the impression I had received from other sources, of the wonderful scenery of California.

Lady Harriet Hervey is an extremely pleasant and amiable person, and seems to have a very good understanding.

Mr. Prior is a very clever and interesting young man, who has lately gained the first of the Natural Science fellowships at Trinity.   He seems likely to be a distinguished man.   The rest of the party, I think I have mentioned before.

———

February, 16th.

My Barton rent audit.   Very satisfactory.

Went on reading Hume on "Morals," and Dunlop's "Spain."

LETTER.

Barton, Bury St, Edmund's.

February 18, 1871.

My Dear Katharine,

I ought sooner to have written to con- 1871.
gratulate you on Leonard's first appearance *in
print* ; an important event in a young man's life.   I
think his letter to *Nature* very good ; modest,
clear and much to the purpose ; free from *verbiage*,
and showing that he well understood his subject.   I
remember being much impressed, long ago, by a
passage in Hare's "Guesses at Truth," to the
effect that the first requisite for good writing is to
know precisely what you want to say.   Charles
Lyell's writings have always been conspicuous for
the observance of this maxim, and Leonard shows
that he also feels the force of it.

I am glad to see that Mr. Carruthers has been
appointed to the head of the botanical department
in the British Museum.

We are enjoying the mild weather after so much
cold.   The yellow aconites and snowdrops are in
profusion ;   the crocuses and hepaticas coming out
beautifully,   and   the   birds   beginning   to   make
themselves heard.

Dear Charles and Mary's visit was a great
pleasure,   and   we   had   a   very   agreeable   party
altogether, but I was grieved to see Charles so far
from well.   I was much pleased with your Frank.

The newspapers—the debates in Parliament—
now take up a great deal of one's reading time, and

1871. altogether, though we are now quite alone, and there is no particularly pressing business, I never find that I get through half what I intended to do.

I am, however, reading David Hume's "Enquiry concerning the Principles of Morals," which is very good. His style appears to me perfect for that sort of subject; more suited for such than for history. I have also got the second volume of Stevenson's "Birds of Norfolk" which is full of entertaining and (to a naturalist) interesting matter. I much admired Mr. Auberon Herbert's speech in the House of Commons the night before last.

We are both well I am thankful to say. Poor Palmer is better but makes very slow progress.

Believe me ever,

Your affectionate brother,

CHARLES J. F. BUNBURY.

P.S.—Fanny's Zoé is quite well. I see an account of a new book called "The Paradise of Birds." Do you not think that Barton might almost deserve that title?

JOURNAL.

February 18th.

Walked with Fanny and little Arthur and her dog Boy. A visit from Lady Bristol and Lady Augustus Hervey.

Read I. Corinth., ch. 12., in Greek, with Stanley's notes. Read in evening, Kingsley's fine sermon on The Charity of God.

February 23rd.

A walk about the park with Fanny, two dogs and the fawn.

LETTER.

Barton, Bury St. Edmund's
February 25th, 1871.

My dear Joanna,

This morning we have been reading (what Leonora sent us), your letter to Annie, containing a most agreeable and interesting account of all you have been seeing at Rome. I can quite enter into all your feelings about it ; there is certainly nothing so interesting as Rome : I have been there three times, each time staying two months or more, and I should be very glad to go again. All seems as familiar to me, and rises up as vividly before my eyes when I think of it, as the neighbourhood of Bury. But in a first visit, especially a short one, the impression is apt to be a little bewildering and overpowering. You will enjoy it more the second time. What I enjoyed most, I think, and what remain most delightfully impressed on my mind, are not so much the special *sights* as the general effect of the ancient and now half-deserted districts

1871. of Rome, the Forum, and from thence onwards to
the Coliseum, the ancient walls and the surrounding
country. There is something magical, indescribable
in the views from the Palatine, the Janiculum on
the tops of Caracallas Baths, over the Campagna to
the Alban and the Sabine mountains.

Fanny's journal will have given you full information about our two very agreeable parties in the
last month and the first half of this. They were
parties of the sort which leave a lasting impression
of pleasure on the mind, which have no drawback
at the time, except that they are so soon over, and
to which one looks back with unmixed satisfaction.

As to books, my reading since I last wrote to you
has been chiefly among not very new ones, with
the exception, indeed, of the " Students' Elements
of Geology," which you have doubtless read and
admired. Mackintosh's " Dissertation " led me to
read Shaftesbury's (the third Lord Shaftesbury's)
" Enquiry concerning Virtue," but I was disappointed, I did not feel that I learned anything
from it, and the style I thought pompous and heavy.
I am now reading Hume on the " Principles of
Morals," which I like very much. By the way, I
lent Mackintosh's " Dissertation " to Sarah Hervey,
and she is delighted with it : she has a very fine
mind, and a great taste for moral philosophy. In
history, I have read Dunlop's " Memoirs of Spain
under Philip IV. and Charles II. :" a clear, simple,
and fairly well written sketch of a portion of
history of which I knew very little. I am now
engaged with Lord Mahon's (Stanhope's) " War of

the Succession in Spain," which is very pleasant
reading: this joins on to Dunlop, and also connects
itself with the "Reign of Queen Anne," which I
read last year: and after this I think of reading
some of the French Memoirs relating to the same
time.

Our snowdrops, crocuses (of several species),
hepaticas and blue squills are now in great profusion
and beauty. I do not as yet perceive that the
severe winter has done any very serious damage to
the trees or shrubs: but it is too early yet to
speak with certainty.

I have no heart to write much about France: it
is too melancholy. Briefly, I will say that I agree
to a great extent with you and Susan, though I do
not go quite so far. Fanny *does*, I think, go quite
as far, and is in a terrible state of mind about it.
I rejoice in the probability of Peace, because I
think there is no hope of success for France, and a
prolongation of the war would be only a prolongation
of murder and misery: but I grieve that France
should have to submit to such oppressive and
iniquitous terms. I especially detest the Prussians
for their arrogant determiation to march into Paris.
Their insisting on this involves a frightful danger:
for the act of one fanatical fool or one ruffian might
bring on a struggle of which the consequences might
be too horrible to contemplate.

I agree with Susan, that England is by no means
safe, and this, because our Ministers cannot make
up their minds to contemplate seriously the con-
tingency of war. The Gladstonites have always

1871. sought popularity, and no Ministry can be popular in England that does not reduce taxation. John Bull is terribly sensitive in the region of the pocket, and any twinge in that organ, makes him immediately roar out for retrenchment.

With much love to Susan,

Believe me ever,

Your affectionate brother,

CHARLES J. F. BUNBURY.

Barton, Bury St. Edmund's
February, 26th, 1871.

My Dear Mary,

I return with many thanks Colonel Smyth's pamphlet, in which I have found a good deal to interest me, about the battles and the Prussian tactics, though, as was to be expected, there is also much which is too technical for me, requiring a knowledge of the details of artillery and fortification. There is no doubt that the Prussian military force is the most formidable of which we have any record in history, as numerous as the Tartars or Huns, and as highly disciplined as the ancient Romans, therefore well worthy to be studied by all military men. And as it seems evident that for a long time to come, all European States must make the preparation for war their chief end and aim : the Prussian system seems to be the best model to follow. I am afraid our Government are by no means sufficiently alive to this necessity. Mr. Cardwell's speech was a very good speech, but

his plan, as regards the militia, seems very 1871.
insufficient.

I see in the *Times* that the Jardin des Plantes has
been almost entirely ruined by the bombardment.
It is a mercy and a marvel that Decaisne and
Milne Edwards were not killed.

I have had an enormously long paper (55 closely
written folio pages) referred to me by the Royal
Society to report upon ;—Professor Williamson on
the Structure of Calamites : and am slowly toiling
through it. It is most minute and elaborate and
full of matter, but a heavy paper in every sense.

Did Lyell remark a short notice in *Nature* of
the 2nd of this month on the *Antholites* of the Coal
formation ? A subject on which I know he used to
be curious.

It is stated that a Mr. Peach exhibited at
Edinburgh a specimen of *Antholites Pitcairniæ*, with
its fruit (Cardiocarpon) attached, from the Coal
field near Falkirk. If this be accurate, it throws a
great light on the structure. It is quite in accordance with what Joseph Hooker suggested to me
long ago—that the Antholites might be a fruit-
bearing *spadix* analogous to that of some Palms, or
possibly of Cycas.

Believe me ever your very affectionate brother,

CHARLES J. F. BUNBURY.

JOURNAL.

1871.   Scott's report of the Rent Audit at Mildenhall—
very satisfactory.

Received an excellent copy, in fine condition, of
Greville's "Cryptogamic Flora," 6 volumes, bought
from Mitchell in Parliament Street.

Mrs. George Jones, the widow of my old friend
the artist, has lately, very kindly and generously,
made me a present of several original and very
characteristic drawings of my grandfather, which
had long been in her husband's possession.

———

There are now good hopes, but by no means a
certainty, of Peace.   The preliminaries have indeed
been signed, but they have not yet been ratified by
the National Assembly.   The conditions of peace
are altogether most oppressive and iniquitous ; yet,
if the Asssmbly be rash enough to reject them, the
renewed struggle will be more fiendish than ever,
and utterly hopeless—at least so it appears to me.

———

Walked with Fanny, the dogs, and the fawn.

———

A beautiful day.   Captain Horton and the younger

Miss Broke* called, and we walked round the 1871. grounds with them.

———

March 3rd.

A lovely and delicious day—perfect spring. We enjoyed a stroll about the park.

Yesterday's *Times* brought the good news that the National Assembly at Bordeaux have ratified the preliminaries of Peace. So, *for the present*, there will be no more horrors. The Prussian army, or part of it at least, had marched into Paris, and no mischief had as yet happened ; but one cannot feel comfortable till they are *out* again. And how long will the peace last ? At the utmost, only till the French have recovered strength enough to take their revenge.

———

March 4th.

Sent off Prof. Williamson's paper to the Royal Society.

We may now hope that the war is really at an end, for the ratifications have been exchanged, and the German army has actually evacuated Paris.

I am most particularly glad to learn (from *Nature* of March 2nd), that the Museum of Natural History at Strasbourg has escaped uninjured from the bombardment.—"The fine collections of mammals, of birds, and of fossils, the result of many years of labour of Prof. Schimper, are perfectly untouched."

———

* Now Lady Loraine.

1871.   March 5th.

We went to morning church, and received the
Sacrament.

March 8th.

Business with Scott;—the Mildenhall School
Board.

March 10th.

Visit from Lady Augustus and Lord John Hervey.

March 11th.

Lady Hoste, Miss Horton and Miss Broke, and
Mr. Murray came to luncheon with us, and several
children came also to play with Arthur MacMurdo,
whose birthday was celebrated to-day.

March 14th.

We went to Hardwick and spent the afternoon
with Lady Cullum; saw her hothouses, &c. She
was very pleasant.

March 15th.

Business with Scott—Mildenhall School Board.
Mr. Paine came with the Churchwardens'
accounts, and I went through them with him and
Scott.

Lady Cullum shewed us yesterday at Hardwick a
tree of the Cornelian Cherry, Cornus mascula,
perfectly covered with its yellow flowers, which come

out before any leaves, so as to have a very bright 1871.
and gay appearance. This is the first time I remember to have seen it in a garden, though it seems
to have been frequently cultivated by our ancestors.
I have seen it in flower in a wild state on the
Apennines, between Spoleto and Terui; but there
it grew as a shrub, whereas Lady Cullum's is a well-
formed tree, perhaps 20ft. high.

LETTER.

Barton,
March 16th, 1871.

My Dear Katharine,

I am very glad to hear that you have got a
a specimen of so very rare a plant as *Lobelia urens;*
when we are in town, I shall hope to see your
specimen, as it is a plant I only know by description. It is quite a *western* species, and not found in
Italy, I believe. I thought it was extinct in
England.

I hope the ferns we sent you will thrive. Ours
(in the houses) are mostly in fine condition, but the
" gold ferns " of which we used to have some very
fine plants, seem to have departed this life.

The conservatory is in great beauty, with
camelias, azaleas, a fine hybrid rhododendron,
begonias, scarlet and yellow tulips, scarlet salvias,
forget-me-nots, roses, Calla Æthiopica, and some
other things; and in the little hot-house we have
a scarlet passion flower, P. racemosa.

1871.   I have lately been looking through the last report of Dr. Carpenter and Mr. Gwyn Jeffreys on "Deep Sea Dredging." Have you seen it? Their field of operation this time was partly off the coast of Portugal, and about the Straits of Gibraltar, partly in the Mediterranean.

They do not seem to have made such remarkable discoveries as in their former cruises; but what is curious, they quite confirmed Edward Forbes's observations as to the absence of animal life in the Mediterranean at considerable depths, though they had found abundance and variety af animals in the Atlantic at greater depths.

We are both well, though we were re-vaccinated the day before yesterday; but the effects are not yet beginning to be perceived.

Fanny is reading Lord Brougham's "Autobiography," and seems to find it very interesting. I wait till she has finished the volume to begin upon it. I wish one could feel a moderate degree of confidence in its truthfulness.

This morning to our great astonishment, we had a heavy fall of snow, and it was not all entirely melted before dark.

Ever your affectionate brother,
CHARLES J. F. BUNBURY.

JOURNAL.

Sunday, March 19th.

We went to morning Church.

Read a good sermon by the Dean of Wells on 1871.
Government of the Tongue.

———

Business with Scott.

The poor little fawn ill.

Terrible news from Paris. The Government
after too long delaying to coerce the refractory
National Guard, at length attempted to act with
vigour, but the Reds broke out into open and
successful rebellion; some troops of the line who
were led against them joined them, and at the
date of the latest accounts they were masters of
Paris.

Two Generals had been most atrociously
murdered, and the members of the Government
had escaped to Versailles. The result will
probably be a civil war in France, and if the
Reds triumph, there will be a repetition of all
the horrors of '93 and '94.

———

Sad news of the dangerous illness of Kate
MacMurdo.

I went to the Quarter Sessions at Bury—little
business.

We dined with the Abrahams. Met the Bishop
of Ely and Mrs. Brown, Lady Cullum, the
Hortons, Chapmans, Bains, etc.

———

Splendid weather. We were out much of the day strolling about the grounds.

The news from Paris worse and worse; a shocking massacre of unarmed citizens by the National Guards in the streets. There seems a sad want of vigour and energy in the Ministers and National Assembly, a disposition to temporize and dally with the Red ruffians, which can lead to no good. If Thiers and his colleagues are really waiting only till they can bring up well-affected troops from the provinces, well and good: otherwise their conduct appears lamentably weak and irresolute.

_____ ___

March 24th.

Long talk with Scott on business.   New heating. Extension of drain.   Clerk to Mildenhall School Board.   Cottage Hospital.
Miss Kinloch arrived.

___ ___

March 25th.

A splendid day; positively hot. Attended the Barton vestry meeting; all very amicable and comfortable.

We (including Miss Kinloch) dined with Lady Cullum — a pleasant party — the Bristols, the Augustus Herveys, Lord John, the Abrahams, the Wilsons.

___ ___

Long talk with Scott on business. Discussion of plan for heating house.

Finished " Memoires de Berwick."

Our dinner party :— Lady Cullum, Lord John Hervey, the Hortons, Mrs. and Miss Wilson, the Victor Paleys, Mr. Abraham.

———

March 28th.

We went to the Bury Athenæum—heard Lord John Hervey's lecture on Bloomfield—very good.

———

March 29th.

A long discussion with Scott and Fanny on plans for new cottages.

———

March 31st.

Visit from Mr. and Mrs. Percy Smith and Charles Murray.

We dined with the Maitland Wilsons—met the Bishop and Mrs. Browne, the Milbanks, Hortons and others—a pleasant party.

———

April 1st.

We went to Ashfield, saw the Milbanks and their pretty garden.

———

April 2nd.

Read an excellent sermon by Mr. Church, on Christian Society.

———

1871.   April 3rd.

Completed and gave in my return for the census. The number of persons who slept in this house last night was 23, of whom only 6 were males.

——— ——

April 4th.

Agnes Kinloch, who has been staying here ten days, left us ; she is an uncommonly pleasing, amiable, interesting girl.

——— ——

April 5th.

We had the comfort of receiving a hopeful account of dear Kate MacMurdo, who is now, we may trust, out of present danger.

Mrs. Byrne arrived.

——— ——

April 6th.

The civil war in France between the Reds in Paris and the Government at Versailles has begun in earnest ; and happily the Reds have had much the worst of it in the first encounters ; the regular troops have behaved well, and shown steadiness, and as more of the soldiers, who were prisoners in Germany return to the army, the party of order will gain more and more.

——— ——

April 8th.

In the course of this week, I have read through Mr. Goschen's speech, expounding his great scheme of local taxation and local government. It is a large, courageous, and important measure, and

appears to me (as far as I can yet judge) a compre- 1871.
hensive and statesmanlike one.*

We went to morning Church and received the
Sacrament.

Lady Cullum and two old Miss Boltons came to
afternoon tea.

I learn from the census enumerators that the
population of this parish has increased by 30 since
1861, the number now being 878. From 1851 to
1861, there was a *decrease* of 9.

The 27th anniversary of our happy engagement.
Lady Cullum dined with us.

Henry writes to me respecting Mr. Cardwell's
Army Reform Bill :—

" It seems to me that Government have begun at
the wrong end. When guns, horses and men were
wanted, they have begun by upsetting the old
regimental system as far as they could, by unsettling
the position of officers and disgusting them. Every
one but the Radicals knew that the only part of our
army system that was really good, was the regimental system. Our regiments were nearly perfect,
in spite of the pains taken on service to spoil them
by withdrawing for employment on the staff or in

---

* This bill has been withdrawn. (May 12th).

1871. the engineer department, every man who knew any trade, or could make himself useful. Our supply departments have always failed hitherto, and probably will continue to do so, at least at the beginning of every war.

With regard to the purchase system ; no one can defend it in theory, but in practice it is both convenient and economical to the country. Its abolition will entail not only a heavy immediate charge, but a great permanent tax in the shape of retirement for old officers. Under the purchase system such retirement was not wanted. When an officer was tired of the service, he realized the value of his commission and there was an end to the matter. In the case of an officer who was poor, or perhaps had a large family, he would say he could not afford to go unless a certain sum was made up, sufficient for him to live upon hereafter. That sum would be raised in the regiment, and the country paid nothing. In future such a man must remain on in the service, grumbling and discontented, weary of the monotony of barrack life, without any hope, until he reaches the age at which he will be compelled to retire, and then the country must pension him. I cannot see why it should be assumed that because a man has money enough to purchase his commission he must necessarily be a bigger fool than the poor man, who cannot purchase. It seems to me that the richer man is likely to have had a superior education, and opportunities of acquiring knowledge that his poorer neighbour could not possess. Then, again,

it by no means followed that the man with money 1871.
could always work his way to the top very quickly.
If unfavourably reported on, his promotion would
be stopped, or he would be made to leave his
regiment. There were flagrant instances of men
being pushed up unfairly, such as Lord Cardigan,
who became a Lieut.-Colonel in six years. But
that might also occur under a system of selection.
I fear we shall get an inferior class of men into the
service, and that regiments will suffer in consequence. Soldiers like to be commanded by
gentlemen; they hate officers of their own class."

The article in this month's *Fraser* on Army
Reform (by Col. Macdougal, I understand) is very
good; very unfavourable to the Government
scheme.

LETTER.

Barton, Bury St. Edmund's
April 14th, '71.

My Dear Katharine,

Miss Kinloch's visit was very pleasant to
us: she is an uncommonly interesting girl. The
last time she was here, before this, she and Sarah
Napier and Kate MacMurdo struck up a special
friendship together.

I wish you could see the conservatory, it is so
crammed full of beautiful flowers—roses, camellias,
azaleas, scarlet salvias, primulas, cortusoides, tro-
pæolum tricolor, dielytra, begonias, æschynanthus,

1871. &c. Fanny's new "spring garden," too, in the front, is very gay now with hyacinths of many colours, tulips, anemones, primroses, polyanthus, forget-me-nots. The lovely delicate green of the young leaves is beginning to show itself everywhere, and the birds are joyfully busy. Every year I feel fresh and undiminished enjoyment in the beauty of spring.

Little Arthur MacMurdo, who is a great discoverer of birds' nests, has found a *water-hen's* nest *on a tree*, near the little pond in the pleasure ground. —I was at first incredulous, but I find in Stevenson's "Birds of Norfolk" that there are several instances recorded of that bird nesting on trees,—in one instance as much as 20 feet above the ground.

It is very pleasant to hear of Charles and Mary making such an agreeable and interesting tour, and enjoying it so much.

Lord Brougham's Memoir, I must confess, interested me but very moderately, and did not raise my opinion of him. How inferior in interest to the Memoirs of your uncle, of Romilly and Mackintosh ! Your aunt, Mrs. Byrne, says that his account of the foundation of the *Edinburgh Review* is inaccurate ; I cannot judge between them.

I am now reading the Memoirs of St. Simon, which are connected with the course of historical reading I have followed for some time past. Lord Stanhope's "Queen Anne," and "War of the Spanish Succession," Dunlop's "Spain" under Philip IV. and Charles II., Memoirs of Marshal

Berwick, and now St. Simon. I am also reading 1871.
Cicero on " Morals "—De officiis—which is very
noble. I had read it before, but more than 20 years
ago.

I read a pretty little novel called *Vera*, recommended to me by Miss Kinloch, who is acquainted
with the authoress : the heroine is a Russian girl,
and is charming.

What a dreadful state of things in Paris ! more
hopeless, I think, than in the siege. I am so sorry
for the Tourgueneffs. I am very sorry too for Mrs.
Byrne, who is naturally alarmed for the safety of
her apartments and her personal property.

We are both quite well, I am thankful to say. I
hope you have very good accounts of Leonard and
Frank.

Believe me ever,

Your affectionate brother,

CHARLES J. F. BUNBURY.

JOURNAL.

April 15th.

Mrs. Wilson and Captain Horton spent all the
afternoon with us—their little boys coming to play
with Arthur.

April 16th.

Read a good sermon by Mr. Church, on Civilization and Religion.

1871.　A much more favourable account, thank God, of dear Kate MacMurdo.

————

Much rain and some thunder.　Our dinner party : —the Abrahams, Lady Cullum, the Hortons, the M. Wilsons, H. Wilsons, the Percy Smiths, Murray, Mr. Montgomery.

————

Mrs. Percy Smith told me yesterday, that she had that day seen two swallows, and had heard the nightingale.

————

The civil war around Paris is still going on with deplorable waste of life and property, and no perceptible approach to a conclusion in any way.　The Reds can gain no ground in the direction of Versailles, and the Government forces only bombard Paris, without being able even to advance up to the enceinte, much less to enter it.　The state of things is incomprehensible and depressing.

————

We dined with the Hortons—met the Bishop of Ely and Mrs. Browne, the Maharajah and Maharanee, the Archdeacon and Mrs. Chapman, Lady Cullum, Patrick Blake, and others.

LETTER.

My dear Joanna,   1871.

I suppose the neighbourhood of Florence
is now in great beauty with the wild spring flowers ;
those of Italy are so very beautiful and varied.
I have never been at Florence just at this time
of year, but the lawns and woods about Rome are
perfect botanical paradises.   Here we have a quantity of spring flowers in the garden, and are
thoroughly enjoying the beauty of spring ; an ever
new pleasure, which I enjoy with ever new zest
every year, and relish quite as much as when I was
young. Fanny laid out last autumn a new spring
garden on the lawn, in front of the conservatory, and
it is now very gay and pretty indeed ; full of
hyacinths, tulips of many colours, anemones, primroses, polyanthuses and forget-me-nots. The conservatory is also in great beauty. I have been much
interested by Susan's remarks (which Fanny has
read to me) on the Memoirs of Mackintosh. It
is one of my favourite books ; one which I recur to
again and again ; and Susan's observations have led
me to refresh my recollection of it. His Dissertation
on Ethical Science (which I have lent to Sarah
Hervey), is also a great favourite with me. Of
all the modern Memoirs, that which I admire most
of all is Romilly's : the fragment of autobiography
with which it begins has always appeared to me one

1871. of the most beautiful and interesting things I have
ever read. (By the "Modern Memoirs," I mean
those made up chiefly of the journals and letters of
the subjects of them—not biographies of the old
kind). The memoirs of Lord Brougham, as far
as they go, have disappointed me much. I must
confess I feel no confidence in his accuracy, and
he is not entertaining, which I did expect he would
be.

There is an excellent article in the new number of
the *Quarterly*, which I think would interest you,—
on the Usages of War; showing the necessity of
civilized nations seriously considering and agreeing
upon improvements in the laws of war, so as to
prevent for the future such cruelties as were lately
committed by the Prussians. The civil war at Paris
is a dreadful calamity, and the state of France
appears to me ever worse and more hopeless than
it was during the late siege. Thiers is placed in
the midst of difficulties, which even a man of the
very highest order could hardly overcome; and he
is, I am afraid, by no means a first-rate man.
Indeed one of the very bad signs for France is the
non-appearance of any man of conspicuous and
commanding ability. It is not yet I am afraid by
any means clear how this present struggle may end;
and if the Commune are victorious over Macmahon,
then will come a war throughout France between
the cities and the country—horrors worse than those
of '93, and in the end, again a despotism.

It is curious that the Jacobins of '93 put forth
"La Republique une et indivisible," as one of their

principal watchwords ; and the Reds of '71, on the 1871.
contrary, seek to break up France into a Federation
of independent cities.   In the meantime it is a most
melancholy crisis for all who are interested in Paris.

You will have heard all about ourselves from
Fanny.   She is quite well, and her heart is shared
between Arthur MacMurdo and the paroquet Zoé.

With much love to dear Susan,

Believe me,
Ever your affectionate brother,
CHARLES J. F. BUNBURY.

JOURNAL.

April 25th.

Walked to the Cage Grove on Lofts' farm.
Studied Mosses.   Heard the cuckoo for the first
time this year.

April 26th.

Mrs. Byrne went away.

We went, with Arthur with us in the carriage,
and Herbert riding, to the Shrub, and then to the
Heath : enjoyed the beauty of the spring and the
wild flowers and the nightingale's song.

April 28th.

Wrote to Sarah Hervey.   Delighted at Mr.
Lowe's discomfiture and surrender of his detestable
Budget,—the worst Budget, I think, that I remem-
ber in my time.

April 29th.

Read some more of the *Daily News* War Correspondence, and began to read Galton on Hereditary Genius.

_____

April 30th.

I see announced in the *Gardeners' Chronicle*, and in *Nature*, the death of Mr. William Wilson, the great authority on British Mosses, and great in British Mosses altogether.

_____

May 1st.

We went to Mildenhall. Herbert and young Murray joined us there.

_____

May 2nd.

Mildenhall. We went through several of the plantations with Scott and Betts, Herbert and Murray. Visited Julia Betts at the Nursery. Inspected the new plantations—well satisfied. Herbert and Murray went away.

_____

May 3rd.

Mildenhall. A wet day. We visited Mr. Lott. I visited Mr. Isaacson and the boys' school. Fanny went to see West Row. Packed some specimens.

_____

May 4th.

We went with Scott through some more of the plantations. Mr. Lott came to luncheon with us.

_____

May 5th.

We returned home from Mildenhall, having gone through most of the plantations, inspected the new

ones in particular, and visited Julia Betts and her 1871.
husband at the Nursery Cottage, also visited the
schools.

Edward writes to me :—" It is strange what a
" position the Gladstone Ministry have fallen into
" altogether. He is said to be still popular with the
" constituencies, and that he would still carry all
" before him in case of an election, but he has
" certainly lost the command of the House of
" Commons, and to a still greater degree of the
" upper classes out of it. It is not expected that he
" will carry any of the important measures that he
" has brought in this session."

———

May 8th.

We went to Stutton, to our dear old friends the
Millses, whom we found well and cheerful, and as
kind, cordial, and pleasant as ever. John Hervey
was staying in the house, and Mr. Harman, a
nephew of Mr. Mills.

———

May 9th.

Mr. Harman went away early. I walked about
the grounds with Mr. Mills in morning, in afternoon
Mrs. Mills took us in the carriage to Woolverstone,
where we saw the new dairy and the garden. Mr.
and Mrs. Berners were away. Mr. William
Gurdon, Mr. Ambrose* and Mr. Woodd came to
dinner.

* A retired Solicitor, and a very agreeable man.

1871.   May 10th.

We went out driving with the Millses, round by
Tattingstone to Wherstead.   Weather dull and
dreary, with a bitter E. wind.

———

May 11th.

We returned home : all well, thank God.

A very fine bright day, with a very cold E. wind.
Read the debate in the House of Commons on Mr.
Miall's motion for the overthrow of the English
Church.   Sir Roundell Palmer's speech, noble—
admirable.   The majority is good, but I am sorry
that Mr. Miall got so many as 89 to vote with him.

———

May 12th.

Mr. and Mrs. Montgomerie came to luncheon—
she very pretty.

Grieved to learn that Lady Adair is dead :—an
amiable, agreeable, excellent person, whom we have
long known.   She has been in very bad health for
more than a year past, and I suppose this bitter
season has brought the end sooner than was
expected.

I am sorry to see in the papers the death of Sir
John Herschel : not that it was premature, or could
be altogether unexpected, for he had completed 79
years, and had long looked very feeble : but one
cannot help feeling regret when one of the patriarchs
of science, so wise and so good a man, is taken
away.   His was a happy, an honourable, a noble
career.   I am very glad that we saw him last year ;

we visited him in his beautiful home at Collingwood, 1871.
he was looking *very* old in the face, but did not
seem very infirm in body, and his mind was
evidently quite clear, active and vigorous. In that
beautiful home he has died.

Mary Lyell has sent us a letter she has received
from Mrs. Gordon, giving an account of Sir John
Herschel's death. It was as easy and peaceful as
his best friends could wish. He had been very
feeble for some time before, but no danger was
apprehended. In the morning his family observed that he slept later than usual, and sent
for the doctor, who came, and found the breath
just departing. So, without a word or a struggle,
he passed from sleep into death. What a suitable
and happy close to such a life.

The Commune of Paris have destroyed the
Column of the Place Vendome. What an absurd
piece of barbarism !

Up to London, to the St. Pancras station of the
Great Eastern : a very good station, and well
situated ; an immense improvement upon Shore-
ditch. Maitland Wilson in the same compartment
with us. Clement and Herbert dined with us.

We went to morning service at St. Michael's, Chester Square, and heard a charity sermon, rather striking and impressive, from the Archbishop of York.

In the afternoon, the Lyells, Edward and the Bowyers came to see us. Charles and Mary both looking well; the latter astonishingly young and handsome. They talked with great delight of their late tour in Devon and Cornwall. They had however, been prevented by bad weather from seeing the Lizard and Kynance Cove, but had been mnch impressed by the scenery of the Land's End. Lyell told me that, in a geological view, anything in this tour had interested him more than the Blackdown beds, which are certainly upper cretaceous (upper greensand, he says, and probably corresponding also in part to the gault) forming a flat table-land, and lying horizontally and un-disturbed immediately on the upper Trias, which is also nearly horizontal; the whole of the Liassic and Oolitic series being absent.

A visit from George Napier.

At last that most strange affair, the French siege of Paris, seems to be approaching its end. The telegrams in this morning's *Times* announce that the Government troops have entered Paris, having gained possession of two of the gates, seemingly without close fighting, having over-powered all

resistance by their superior artillery fire. It still 1871.
remains to be seen whether the Reds will, as they
have so long threatened, defend their barricades
and fight from street to street. If they do, the
struggle may yet be long and bloody, and perhaps
even the result doubtful.

May 23rd.

Lady Campbell brought us the *very* welcome
news that dear Kate MacMurdo has arrived safe at
Simla, and though extremely delicate, seems to have
quite recovered from the fever.

May 24th.

A beautiful day. Went out in open carriage with
Fanny, Cissy and Emmy — we went through
Battersea Park.

Visits from George, Cecil, Mary Lyell, Lady
Cullum, Col. and Miss Kinloch, the Miss Moores,
Catty Napier, Lady Rayleigh and Miss Strutt.

Cecil dined with us.

The Government troops in Paris have, by the last
accounts, gained a success which must be decisive ;
they have got possession of the Heights of Mont-
martre, the very citadel of the Red insurrection.

May 25th.

Dreadful news from Paris. The Reds, in their
fury and despair, have deliberately set fire to the
city, and in particular, to the principal public
buildings. The Tuileries are "a heap of ashes ;"

U

1871. there is little hope of saving the Louvre; and it is feared that Notre Dame, the Palais de Justice, and the Sainte Chapelle, the most interesting relics of mediæval Paris were also on fire. The description by the *Times* correspondent, of the scene—enormous fires raging in several quarters at once, is horribly interesting. It would hardly be possible to calculate the loss, not to France only, but to the whole civilized world.

I am very glad to hear from Bentham that Adolphe Brongniart is safe out of Paris, though in bad health. He was in Paris all through the first (the Prussian) siege, and his health was very much injured by it; he is now living on a small property he has in Normandy. Poor Decaisne, however, is in Paris.

—————

May 26th.

Went out driving with Fanny—we paid many visits; saw Mrs. Hugh Adair, Mr. and Lady Anne Lloyd, the Ladies Legge, Mrs. Young. Miss Kinloch and her brother, Minnie and Sarah, Cissy and Emmy, Edward and Clement dined with us.

Happily, by this day's accounts, the mischief done appears to be not quite so great as was feared. The greater part of the Louvre is said to have been saved; the Palais de Justice is said to have escaped, and the Sainte Chapelle is but slightly damaged. The Palais Royal and the Hôtel de Ville are partly destroyed. The fighting is still going on, and attempts are still made to renew and extend the conflagrations.

May 27th.   1871.

The news from Paris is really too frightful to dwell upon. The struggle still goes on, though utterly desperate on the part of the Reds, and the horrors which are accumulated on the wretched city seem like the most awful tragedies that one has read of in history.

The civil war in Paris has become so ferocious, and shows such a depth of hatred that one does not see how real peace, and an orderly and well regulated society are ever to be re-established. The German invasion was a mild misfortune to France, in comparison with this.

————

May 29th.

The accounts from Paris continue to be horribly interesting. The crimes of the Red ruffians have culminated in the murder of the good Archbishop and a great many other of what they chose to call hostages. They are now driven out of all their principal strongholds, and closely hemmed in, and we may hope that the dreadful struggle is nearly over. There seems to be no doubt that their intention was to destroy Paris entirely, and they have succeeded in destroying a great deal, in doing infinitely more mischief than France has ever suffered from foreign enemies. They have succeeded also in destroying all possibility of political liberty in France for a long time to come.

The evening papers bring news that the rebellion is finally crushed, at least as to all open resistance ; and fighting has ceased. But when will peace, order and security be restored to Paris ?

1871.   This afternoon we were very happy to learn from
a copy of a letter lent to us by Mary, that Fanny's
old friend Mme. Tourgueneff and her family are
alive, and that their fine house in the Rue de Lille
has escaped destruction—almost as if by a miracle,
for much the greatest part of that street has been
burnt.   They have lived through a terrible time,
having been in Paris all through this second siege.

May 30th.

The 27th anniversary of our happy marriage—
God be thanked.

Drove out with Fanny; we paid a great number
of visits, beginning with General Fox, but saw no
one except Lady Gardiner and Colonel Lynedoch
Gardiner.   Cissy and Emmy dined with us.

May 31st.

Dear Sarah Hervey arrived, looking well, and
as charming as ever.

June 1st.

Looked into the Royal Academy Exhibition, but
had only half-an-hour to spare for it, so only looked
hastily at a few things.   *Faed's* " A wee bit
fractious" pleased me best in this cursory view.
Millais' " Moses," struck me as grand : and
Leighton's "Alcestis " as excessively strange.

Lady Cullum, Cissy, Emmy and Clement dined
with us.

June 3rd.

Henry came to luncheon, having arrived last
night from Wales.   Lady Louisa and Lady

Wilhelmina Legge also came. We dined with the Hanmers: met Mrs. Shirley, Mrs. Mainwaring, Lord Abingdon, Lord Mostyn, Sir W. Hutt and others. Sir John told me of Lord Mark Kerr's adventures at Paris, which he had just heard from himself. Lord Mark (who is a Major General in our army) was in Paris during these last struggles, and was imprisoned as " suspect " by the Communist authorities, not in one of the regular prisons, but in one of the Municipal buildings, where he was nearly starved. When the Government troops got possession of this building, and found him, they refused to believe that he was a prisoner of the Commune, insisted that he was one of their adherents, and were going to shoot him summarily. In vain he protested that he was an Englishman, and a Major General in H. M. service : they cared nothing for this, and he was in imminent danger, but luckily some of them advised that he should be taken before their superior officer, and that officer fortunately saw the true state of the case and released him.

————

June 4th, Sunday.

Read prayers with Fanny. Visited Charles and Mary Lyell, and had a pleasant talk with them.

Mary Lyell told me that Mr. Grove, the lawyer and philosopher, has been attending closely to the Tichborne case, and thinks it hardly conceivable that the claimant can be an impostor ; the accumulated weight of circumstantial evidence is so great : such a multitude of minute circumstances

1871. about which he has answered correctly, and which it is unintelligible how any impostor could have learned. On the other hand, I am told that the Solicitor-General has (and expresses to his friends) the strongest possible personal conviction that the claimant is an impostor. It is certainly altogether a most extraordinary cause. I hear that all the most practised and eminent lawyers are very much divided in opinion about the Tichborne case. I hear it said that the extreme bitterness which the Solicitor-General displays against the claimant is thought to be injudicious, and calculated to produce a reaction in the minds of the jury. At any rate it is quite clear, from the plaintiff's own evidence on the 5th, that he never ought to be admitted into the society of gentlemen.

Mary Lyell's story about the "Rights of Women:"—an American lady saying, in answer to a question from an English gentleman, that she was quite against the claim of those "rights," because "now in going to and leaving the trains, my "husband carries the carpet-bag, but if we were "exactly on an equality, I should have to carry it "half the way"

June 5th.

Went with Sarah Hervey through the "International" Exhibition at South Kensington: but we had much trouble in finding our way through it, and so wandered on and on without knowing our way: had much fatigue, and did not see much satisfactorily. The only thing we saw which interested

us particularly, was a large model of the gateway of 1871.
the Sauchi Tope, in India, a most remarkable
specimen of architectural sculpture. We looked
into the Albert Hall, which is colossal.

— —

June 6th.

The cold of the weather is still excessive, the
wind continuing always in the N. or N. E. I
hardly ever remember such cold in June. Lady
Napier says she *does*, in 1815. Octavia and
Wilhelmina Legge and Admiral Spencer came to
luncheon. We dined with Mr. Cox and Lady
Wood : met Lord Lyttleton, Mr. Kirkman Hodgson,
&c.

June 7th.

We went to see the Flower Show at the Horticultural Gardens : a very beautiful show of rhododendrons, as usual, and of roses, also some fine
orchids and ferns. I noticed one or two very fine
Lycopodiums, — I suppose Ulicifolium or allied
species; also a beautiful white Cypripedium of the
insigne type (C. Candidum), and another very
curious species C. candadum, with its sepals
hanging down in slender tails to a surprising length.
But the crowd was great, and I could not study the
plants satisfactorily.

William Napier dined with us.

June 8th.

Saw Mrs. Jones, who showed us beautiful sketches

1871. by her husband.   We dined with the Gambier
Parrys: a pleasant party.   I sat by Mrs. Peploe,
sister of our friend Lady Campbell, a very handsome woman, and exceedingly agreeable.   Mr.
Gambier Parry showed us two beautifully carved
ivory diptychs, of Southern French work, of the
13th century.

I hear that the claimant in the Tichborne cause
is so excessively intemperate, drinks such quantities
of brandy and water, that it is thought very likely
he will drink himself to death before the trial is
over.   It is said that one day, during a temporary
retirement from the court, drinking a tumbler of
brandy and water, he said, that was his twelfth that
day.   Lady Doughty and all the real Tichborne
family are in great fear lest he should die before the
cause is decided, for, as he has children, the whole
might have to be tried over again.

June 9th.

Went to see Katharine—had a pleasant talk with
her.   Charles Lyell with her part of the time.

Went to the Zoological Gardens; full of interest
and entertainment, as they always are, but I did
not see any striking novelty.   There is, however,
a fine specimen of the great Ant-eater, and I am
always glad of an opportunity of seeing alive so
very curious and extraordinary an animal ;   so
strange and anomalous, that one can hardly believe
it to be real, even when one sees it walking about.

The Kangaroos also, of which there is a fine col-

lection here, are interesting from their curious 1871.
peculiarities of form and structure ; so are the
Hornbills, of which there are two or three species
in the gardens. In the Monkey house I saw nothing
particular.

In the Parrot house, as usual, most brilliant
and beautiful colouring, and a most horrible discord
of screams ; here, besides the Parrot tribe, were two
or three species of Toucans, with brilliantly-coloured
beaks, and some other curious birds.

Miss Strutt, Mr. Hutchings, and Mr. Medlycott,
dined with us.

———

June 10th.

A visit from Lord John Hervey ; he has just
returned from a visit to Paris, and brought some
photographs to show us. He says that the destruction of the public buildings by fire is not so
complete as might be supposed, that they are reduced to mere *shells* of building, and the roofs have
generally fallen in. In the case of the Tuileries,
the great central cupola has disappeared, but the
garden front is not otherwise, he says, much dis-
figured.

The Hôtel de Ville is more destroyed—the upper
part of it. The destruction, he says, was much less
complete than was expected, owing to the excellent
construction of the buildings ; the fire seldom
spread from one house to another without being
purposely communicated. He went through a great
deal of Paris, but perceived no smell of corpses.
The extreme coldness of the weather, however, has

1871. certainly been, in this respect, an advantage to the
Parisians.   He saw no appearance of excitement in
the people, but he was struck with the general look
of melancholy and absence of animation, and the
comparatively little traffic in the streets.

Fanny and Sarah went to see an amateur play.
Edward dined alone with me.

June 11th, Sunday.

We went to morning church at St. Michael's.
A good charity sermon from Dr. Barry.   Called on
old Sir John Bell; found him flourishing at age
of ninety.

June 12th.

A beautiful and very warm day—the first warm
day of this month,—the wind having at length
changed.

We went to the National Gallery and saw the
pictures lately purchased for it from Sir Robert
Peel, a noble collection.   The beautiful Chapeau de
Paille;   a grand landscape by *De Koniuck* reminding us of *Rubens's* Rainbow Landscape; several
very fine *Ruysdaels;*   a marvellous landscape by
Hobbema, a straight road through a flat country
between miserable lopped trees, with a village at the
end,—a scene ugly in itself, but a wonder of skill
and effect in painting ;   "A Poulterer's Shop," by
Gerard Douw, an exquisite specimen of Dutch
miniature painting ; *Vandyck's* Portrait by himself;
the grand, thoughtful portrait of Dr. Johnson, by

Sir Joshua Reynolds; Boswell, and Admiral Keppel, 1871. also by *Reynolds;* two Interiors, by *De Hooge.* These are what most impressed me in this first visit.

Henry, Cissy, and Emmie dined with us.

June 13th.

A visit from Sally and her little grandson Cecil. Lady Anne and Mrs. Lloyd, Miss Rodney, Mr. and Captain Medlycott, came to luncheon.

We (Fanny and Sarah and I), went to the Zoological Gardens, in company with Mr. Medlycott and his brother Captain Medlycott, and spent a considerable time there. We saw several things which were new to me. One of the keepers showed us (in the monkey house) some very curious fruit-eating bats, (Pteropus) from (tropical ?) Australia, which were quite tame, ate out of his hand, and allowed him to handle him in any way; there were two young ones born here, and one of the females came down from the upper part of the cage with her young one clinging to her side under her wing. Also two young chimpanzees, kept by themselves in a separate and very hot room, exceedingly tame and *child-like,* with most pathetic countenances. Also a large black Gibbon ape, which howled in an extraordinary style. Another keeper showed us in a different house a little Anteater, a new acquisition and a great novelty, never before seen alive in Europe, a very interesting animal, not perhaps quite so extraordinary in its

1871. general appearance as the great Ant-eater, but similar in the shape of its head and of its legs and feet : the tail prehensile, long and comparatively slender, not bushy as in the other. It seemed very tame and gentle, stood up on its hind legs and played with the keeper, and even allowed Fanny to pat and stroke it without seeming at all annoyed. Another keeper *tried* to show us the owl-parrot of New Zealand (Strigops), and the two species of apteryx from the same country : but these being nocturnal animals were extremely unwilling to appear, and hid themselves as much as they could. The apteryx is in a general view very unlike a bird. The bathing of the two African elephants in the great tank was an extremely amusing sight. It suggested the idea of a *burlesque* of the bathing at Brighton.

Mr. Medlycott seems to be an interesting young man, very intelligent and active-minded ; devoted to the study of natural science, and very well informed in it ; having travelled much, and made (as I am told) large collections of natural history in many countries ; his manners very pleasing.

Captain Medlycott, his brother, has similar tastes and has visited a great variety of distant countries.

The first very rainy day of this month, and I should think of the last two months.*

June 14th.

Sir Frederick and Lady Grey came to luncheon with us—very agreeable as they always are.

* There have since been few days without rain.—(July 30th.)

I visited the Royal Academy Exhibition ; greatly 1871.
delighted with *Vicat Cole's* landscapes, called
"Autumn Gold," prefer it decidedly to *Millais'*
"Chill October," which is opposite to it, though
this also is very good.

We dined with the Charles Lyells, met Sir
Edward Ryan, Mr. Duckworth, Mr. Clark (Trinity
College), a Mr. and Mrs. Alford, and one or two
more. In the evening a good many, several
pleasant.

June 15th.

Looked into the Museum of Practical Geology.
There are models of some diamonds, but none of
extraordinary size, from the Cape, but no specimens
of the diamond-bearing deposit of that country.

Rosamond and Arthur Lyell came to luncheon
Cecil and Clement dined with us.

June 16th.

We called on Lady Sidney Osborne (Mrs.
Kingsley's sister, and wife of the celebrated
"S. G. O.") with whom Sarah had become
intimate at Wells. We were charmed with her,
even in this one short visit. She is very like Mrs.
Kingsley in many respects though considerably
older.

Fanny went out to Minnie's. I dined alone with
Sarah.

June 17th.

We (three) dined with the Maitland Wilsons—

1871. met Mr. Ward Hunt, Sir Edward and Lady Gage,
Sir Robert and Lady Buxton, Mr. and Mrs.
Farrer, Henry Wilson.

- - - - -

June 18th.

The Bishop of Bath and Wells, and Admiral
Spenser, came to luncheon. The Bishop stayed
with us some time, delightful as ever.

- - - -

June 19th.

Violent thunderstorms, with furious rain and hail.
It is a very extraordinary season; we have really
had no summer as yet, but one form of bad weather
after another.

Our dinner party; the Bishop of Bath and Wells,
Lord Ross, Lady Mary and Miss Egerton, Lady
Rayleigh and her son and daughter, Mrs. Young,
Sir George Young, Henry and Cissy, Colonel
Yorke, Wiliam Napier, Mr. J. Paley; a pleasant
party. I sat between Lady Mary for whom I have
a great esteem, and Mrs. Young, who is also very
agreeable. Afterwards we had a large evening
party, about 120 altogether.

- - - - -

June 20th.

Visits from Edward and Lady Campbell, Arthur
Hervey, Sir G. Young.

We dined with the Rayleighs—met Sir Roundell
and Lady Laura Palmer, Sir C. and Lady Trevel-
yan, Mr. Coore, Mr. and two Miss Balfours.

Sir George Young told us that "The Battle of

Dorking " was written by a nephew of Lord North- 1871.
brooke, a Mr. Dupré, a cousin of his—(Sir G.
Young). Sir Roundell Palmer, whom we met at
Lord Rayleigh's, told Fanny that he had no doubt
the claimant in the Tichborne cause is an impostor.

June 21st.

Mr. Henry Lofts, the London house-agent whom
I employ, told me that the value of house property
and building ground here, and also of landed
property in the country is continually rising. He
lately sold ⅛ of an acre of building ground here, for
£18,000 ; and he knows that Mr. Smith, the M.P.
for Westminster, lately bought an estate in the
country for £130,000 ; though some might be
inclined to wonder (he said) that a man who is
making 15 per cent of his money in London, should
lay it out on land which will produce only 3 per
cent.

June 22nd.

Leopold Powys and Lady Mary, Major and Mrs.
Peploe, our dear Lady Campbell, Henry, Cissy,
Emmy, and Edward, dined with us ; and Minnie
and Sarah came in afterwards ; an exceedingly
pleasant party.

June 23rd.

Not at all well—could not go out to dinner.
Fanny went alone to dine with Sir Edward Ryan.

1871. A farewell visit from Henry.   Visit also from Frank
Lyell.

--- ---

Our dinner party:—

Mrs. Mills (Mr. Mills being unfortunately pre-
vented by ill health from coming), the Bishop of
Bangor, and Mrs. Campbell, Admiral Spencer,
General and Mrs. Ellice, Mr. and Mrs. Longley,
Minnie Napier, George, Cissy, Edward and Cecil
—pleasant.   In the evening, the Miss Egertons,
Sarah Napier, Rachel and Jessie Bruce, John
Herbert, and a few others.

The news has arrived of the long-expected death
of Charles Lyell's poor brother Tom.

We drank tea with Colonel and Lady Blanche
Morris.   Colonel Morris showed us two beautiful
portraits of his grandmother Lady Morris, the one
by *Reynolds*, the other by *Romney*, both very lovely :
very choice specimens of the two artists.   He seems
to be a man of very fine taste, and his very small
house in Cadogan Place is perfectly crowded with
good pictures as well as with works of decorative
art.

A visit from dear Arthur Hervey, just returned to
town.

--- ---

Emily Napier and her two eldest girls and
Colonel Harding came to luncheon.   I visited
Charles Lyell and Mr. Mills, and had a good talk

with both. I am glad to hear from Charles Lyell 1871. that Joseph Hooker has returned safe and well from Morocco, after going through many dangers as well as hardships. He told Lyell that the breeding of the famous Barbary horses has entirely ceased in Morocco : and for this reason :—that whenever a man was found to be rich enough to possess valuable horses, the government of the country immediately seized him, put him in prison, and perhaps tortured him, till he paid an enormous ransom. This speaks enough for the nature of the government.

We dined with Minnie Napier. I sat by Miss Strutt (Clara) who is an uncommonly clever girl, reads a great deal, and with attention and thought, and talks very well :—reads good sound histories, not light or flashy things. She read through Clarendon last winter, and Burton's " History of Scotland:" and her remarks upon both were very good. In fact she has a remarkably cultivated mind.

————

June 29th.

Royal Academy again.—Admired " Among the Fells " (No. 9) by *Cooper:* a flock of sheep in a snow-storm—the effect of the drifting powdery snow particularly good. Much pleased with " Circe," by *B. Riviere* (1156)—not that Circe herself is anything, but the expression of the pigs is admirable ; they are such perfect pigs, and yet with the human character and feelings showing through in a pathetically comical way. The " Betting Ring," A. D. '68, by *Selous* (1173), is an interesting and

1871. striking picture : a Gladiator prepared to go on the Arena, his *lanista* showing him off to a circle of admiring spectators, men and women.   It well illustrates the mixture of cruelty and licentiousness which characterized the public games of those times.   It might serve as an illustration of Whyte Melville's "Gladiators."   *Frith's* "Salon d'Or" at Homburg (158), very clever, in the style of his "Derby Day."   "Peace," by *Armitage* (19)—French peasants of the next generation turning up the relics of a battle field of the late war,—interesting, telling its story well and simply.   "The Battle of the Frogs and Mice," by *H. Bright* (792), delightful. "The Haunt of the Fallow Deer," by *W. Luker* (457), a beautiful, enjoyable landscape.   "By a Sedgy Brook," *F. S. Cooper*, (493), beautiful.

Our   dinner   party:—Lady   Bell,   Sir   George Young, the Matthew Arnolds, the Godfrey Lushingtons,   the   Bowyers,   Mrs.   Byrne,   Cissy,   Rachael Bruce, William Napier, Edward and Clement.   I was   between   Lady   Bell   and   Mrs.   Lushington :   the former   most   gentle   and   amiable,   the   latter   very pleasing, intelligent and agreeable.

June 30th.

We went out in the open carriage : took Sarah to Westminster Abbey to draw there : we saw the Water Colour Exhibition, and visited Charles and Mary.   Cissy and Emmy dined with us.

Weather   very   strange   all   this   month :—   a most inclement June.   The first 12 days very dry,

with a sharp and piercing N. or N. E. wind: 13th 1871.
and 14th very heavy rain, then a few days of close,
muggy, damp weather: the 19th and some following
days thunderstorms, with violent rain and hail,
then cold, damp, unsettled weather: the 28th and
29th really fine and warm, and this last day bright
and treacherous, with hot sun and cold wind.

July 3rd. Monday.

Visited Lady Smith and Lady Rich: the latter
very agreeable.

July 4th, Tuesday.

Arthur MacMurdo arrived.

July 6th, Thursday.

Fanny ill. I dined with the Hugh Berners: met
Mr. Mills, Col. and Mrs. Anstruther, Admiral
Spencer, Admiral Popham and others.

July 7th

Our dinner party: Minnie Napier presided; Sarah,
Mrs. Mills, Sir Charles and Lady Murray, Sir John
and Lady Hanmer, Philip and Pamela Miles,
Arthur Hervey, W. Napier, Leopold and Minnie
Powys, Lord John Hervey, Mr. W. Gurdon.

July 8th.

Edward's birthday. Sarah Hervey and I went to
visit him; afterwards we saw the Water Colour

1871. Exhibition. Visits from W. Napier, Mr. Mills, Mr. Singleton and his children. Fanny ill again in the night.

My Journal has fallen into arrear. In the first days of this month I had little to note, and since then Fanny's illness (an affection of the trachea, from cold), has disheartened me. But thank God she is now recovering. She was not well enough to preside at our own last dinner party on the 7th, but it was otherwise a pleasant one.

Cissy and Emmy left town the next day, and since then our friends and acquaintances have been fast departing, and I am afraid we shall be left almost the last. Our dear Sarah Hervey left us on the 10th, to return with her father to Wells: very *very* sorry I was to part with her, and she leaves a sad gap. She has been a most delightful companion the whole time. If anything *could* raise still higher the love and esteem and admiration which I have long felt for her, the experience of these last five weeks has done so. I am very glad to find that her family think her looking well.

I have not seen or done much in this first fortnight of July. The 10th I went alone to Kew, failed of seeing Dr. Hooker, but spent two hours in the gardens very agreeably, though I did not see so much of novelty as I expected. The collection of tropical ferns did not appear to me to be either so 1871. rich in species or in such fine condition as it was some years ago. But the palms, pandani, and other great endogenous trees in the palm house, are really very grand, and in superb condition. I noticed, in particular, two very fine and striking Musaceæ : the Abyssinian Ensete, with leaves of wonderful size and beauty, far surpassing those of any common banana in dimensions, brilliancy of green, firmness of texture and lustre of surface : and the famous Urania or Ravenala, or "Traveller's Tree," of Madagascar, with its great banner-like leaves on long stalks, arranged with singular and formal regularity in two opposite vertical ranks. In the palm house, great as its height is, the great bamboo now seems cramped for room, and cannot have full justice for its graceful top, which is in contact with the glass roof. In another house, in a great tank, the Nelumbium, and the Victoria with its wonderful leaves, are flourishing superbly, the Nelumbium bearing abundance of fruit, together with many beautiful water lilies of smaller size, and other curious aquatic plants, such as the Pistia. The old original Araucaria (date 1796) appears to be unhurt.

I learn from a notice of Dr. Hooker's Report on the Gardens. that a great many trees, even old trees, and of common kinds, have died from the drought of last year.

We two, with Minnie and Arthur and his governess, went to the British Museum and took a hasty view of several departments.

Fanny's 57th birthday.

Lady Head and her daughter Amabel came to luncheon. We visited Lady Bell.

July 19th, Wednesday.

We looked into the International Exhibition. Lady Head and her daughter Amabel, Minnie, Col. Yorke, Mr. Walrond, Sir W. Boxall, Edward and Clement, dined with us.

July 20th, Thursday.

We called at Lambeth ; had a glimpse of the Archbishop (looking very well), and saw Mrs. Tait.

Dear Minnie and Sarah dined with us—a farewell dinner, as they are going down into Hampshire. Very sorry I am to part with them.

July 22nd, Saturday.

We drove out to that beautiful place Combe Hurst, to visit our old friends Mr. and Mrs. Sam. Smith, and found them looking tolerably well, and as friendly, cordial, and agreeable as ever. We met there Sir Thomas Erskine May. I was very glad to find Mr. Smith and Sir Thomas May also, taking the same view as I do of Gladstone's *coup d'état* on the Abolition of Purchase :—both condemning it as most arbitrary and unconstitutional. Sir Erskine May said, that the old Whigs in the House of Commons dislike it extremely, while the Radicals are wild with delight. He had asked

Lord Dufferin how the Liberal Peers, who usually support the Government, were disposed, and Lord Dufferin answered that he thought they were very *Whiggish;*—i. e. (said Sir Thomas), they still retain some respect for the British Constitution.

July 24th, Monday.

William and Emily Napier, Edward, Cecil and Clement, dined with us.

July 27th, Thursday.

Down to Barton. Our party consisting of ourselves, Arthur MacMurdo and his governess Miss Jennings, men servants and maids; three pet paroquets and three bullfinches. Arrived all safe, and found all well at our dear home, thank God.

July 28th, Friday.

Barton.

The grass is brilliantly green, such a vivid delicate green, as one usually sees only in the very beginning of summer. The hay has just been all carried; it is a very large crop, and thanks to Scott's watchfulness and good management, has been got in in good condition.

July 29th, Saturday.

We drove in the pony carriage, Arthur riding beside us,—visited the family trees, church, church yard, Scott, &c.

August 1st.

We dined with Lady Cullum — met Lord and Lady Gwydyr, the Abrahams, Hortons, Praeds, Beckford Bevan and others.

August 2nd.

The assizes at Bury. I was foreman of grand jury. A light calendar.

Dined with the judges (Cockburn and Byles).

August 4th.

Mr. Alfred Horwood, deputed by the Historical Manuscripts Commission, came here to examine our MSS., and spent a considerable part of the day here, busy with the collection of autograph letters arranged by my father, and also with other MSS., and rare books. He seems a very intelligent and accomplished man.

August 6th.

We went to morning Church and received the Communion. Mr. Smith preached a very good sermon.

We lounged about the arboretum; much of the afternoon with Clement and Arthur.

August 7th.

I detected cones on Abies Menziesii, in the arboretum—the first time it has borne any here. They are very high up, very near the top of the tree.

Mr. Milbank and Lady Susan, and Miss Pitt came in the afternoon and walked about the gardens and arboretum with us, and dined here.— Lady Cullum also dined with us.

———

August 10th.

Fanny went to Bury to visit the workhouse, and brought back Mr. Farnall, the poor-law inspector.

Mrs. Abraham, Captain and Miss Horton, Maitland and Henry Wilson dined with us.

Edward, who is just come from London, tells me it is now certainly known that the author of "The Battle of Dorking" is Col. Chesney, not the Col. Chesney who is so much noted for his lectures on the Waterloo campaign, and his writings on the Prussian military system ; but either a brother or first cousin of his who has been in the Indian service, and holds some high appointment connected with Indian military education. But—what is very odd—it is said that Col. Chesney, wishing to conceal his authorship, induced his friend Mr. Du Pré, to claim to himself the credit of the book.

====

LETTER.

Barton, Bury St. Edmund's,
August 10th, '71.

My Dear Katharine,

I have been intending to write to you for some time past, but the truth is I have been lazy and disinclined to writing of any kind ; reading

1871. seems to suit one better in this delightfully warm weather.

I hope you have been spending your time agreeably at Edinburgh; and I have been very glad to hear of Leonard's successful public appearance in the character of an authority in scientific matters, which must have given you great pleasure. Not that I had any doubt of his success; I fully expect he will be an eminent man of science; but the beginning is always an interesting crisis.

It was a great delight to escape from London, and find ourselves amidst the green and shade and flowers of our home. I have very much enjoyed this fortnight, and especially the last ten days, during which the weather has been quite glorious.

This place is in its beauty; the sunshine has not yet parched up the grass, which when we came down, was most brilliantly and beautifully green, as green as in May; and this heat following on the abundant rains of July, has brought out flowers in great perfection.

I see some mischief done by the winter, but not much—quite trifling compared to the effects of the winter of '60-'61. Pinus insignis killed outright; Benthamia fragifera killed to the ground; the Pampas grass much hurt; these are the chief particulars of damage. Magnolia grandiflora quite unhurt, and flowering well. The Catalpa just coming into beautiful blossom. Abies Menziesii is bearing cones for the first time, but they are terribly high up.

The out-door ferns are very flourishing, particularly (of foreign ones) Struthiopteris Germanica 1871. and Lomaria alpina; and the former bears fertile fronds for the first time. In the conservatory Eucharis Amazonica, that purest of pure white flowers, is in great beauty ; and Passiflora racemosa in flower.

Fanny, I am happy to say, has I think, quite recovered (thanks to country air and the fine weather) from her London illness. This hot, dry sunny weather agrees prodigiously with me. We are going away from home a fortnight hence to visit the Arthur Herveys at Wells. I do not think that any lesser inducement than the desire of seeing *them* would make me leave home again at this season ; but when we are so far away, we shall perhaps go to the seaside for a little while, to Weymouth and Swanage or Bournemouth ; but to be here again certainly, I hope, by the middle of September.

This splendid weather is a grand thing for the harvest, which is beginning everywhere about us ; in this parish, Mr. Paine has begun cutting his wheat, to-day (August 11th) ; and we have every prospect of a fine time for it.

Edward is here, but goes back to London to-day, and starts on the 22nd for the Black Sea. Arthur is well, and so are the feathered pets.

I am much interested by Nassau senior's " Journals in France and Italy, from 1848 to 1852."

Believe me,
Ever your affectionate brother,
CHARLES J. F. BUNBURY.

LETTER.

Barton, Bury St, Edmund's,
August 12th, 1871.

My dear Mary,

1871. I have read with very great pleasure your interesting letter to Fanny, giving us such agreeable accounts of the tour you have been making in Derbyshire and the North, which must have been really delightful. I am most truly glad that you have been enjoying yourselves so much, and above all that Charles Lyell has improved so much in health. I trust he will continue to gain ground, till you are free from all anxiety about him; that you will both continue to enjoy thoroughly your tour, and return home after it in good health. If you have such glorious weather as we have here, it must be an extraordinary advantage in the beautiful scenery of the Lake country. Here it is quite delightful! too hot for one to go out comfortably in the middle of the day, but most enjoyable when the sun gets low, or in the shade of the trees, which are thoroughly appreciated now. It is glorious weather for the harvest. Fanny has written to you all our incidents (such as they are), and all our plans; and the chief thing I have to tell you is, that I am labouring through a long, cumbrous, ill-written, but very important paper by Prof. Williamson on Coal Plants, which has been referred to me by the Royal Society. It is the second paper; the first, on which I reported last March, was on Calamites, this on Lepidodendron and Sigillaria. What is sent to me is the rough copy—very rough copy—full of erasures and interlineations, so that 1871. it is troublesome to read ; but it is full of matter, a perfect mine of curious and elaborate microscopical observations, illustrated by valuable drawings. He comes to the conclusion that Sigillaria is closely allied to Lepidodendron, and that *both* were crypto-gamous, a conclusion at which Carruthers has also arrived on independent grounds, and in which I fully believe they are right. He is however at variance in this with Adolphe Brogniart, as well as with Prof. Dawson. But though Lepidodendron, &c. were cryptogamous, they had (according to Williamson), a much more highly developed structure in the stem than any existing cryptograms (except tree ferns), and in fact a structure analogous in many respects to that of exogens. There is a very good abstract of this paper in the last part of the Proceedings of the Royal Society. I thought Lyell would like to know this.

We have just heard of a sad thing that happened yesterday at Stowmarket—the explosion of a gun-cotton factory, with the loss (it is reported) of at least 20 lives, and among them the two chief partners or managers of the concern. The explo-sion was heard distinctly here, and was thought to be a clap of thunder, though there was not a cloud in the sky.

Much love from both of us to Lyell and yourself, and sincere thankfulness for the good accounts of both your healths.

Believe me ever your affectionate brother,

CHARLES J. F. BUNBURY.

JOURNAL.

1871. A pleasant drive with Fanny; we visited Mrs. Henry Hardcastle* at Nether Hall, and Mrs. Rickards at Thurston.

———

Read some of Canon Cook's Commentary on Exodus. Mrs. H. Hardcastle and two of her friends, Lady Hoste and Mr. Greene, Archdeacon and Mrs. Chapman came to afternoon tea with us: we showed them the arboretum, &c.

———

Fanny, with Arthur and his governess, went over to Mildenhall and returned late. Examined the flowers of Allamanda neriifolia, which has flowered in one of the houses, and made a note.

———

Read some more of the Commentary on Exodus. Lady Cullum came, and we went together to the Milbanks at Ashfield—admired their garden; we met there the Hortons, Lady Hoste and Mr. Greene, Patrick Blake and the James Blakes.

———

Walked along the green ride, and saw the harvest going on, and the gleaning.

* Daughter of Sir John Herschell.

Arthur went back to school. Up to London.

———— --

48, Eaton Place.

Emily Napier came to luncheon with us:—afterwards we drove out : visited the Miss Richardsons. Emily and William dined with us.

————

To Wells by the 9.15 train from Paddington: arrived at the Palace at Wells about 2 : received with a warm welcome by our dear friends. Lady Augusta Seymour, Lord Arthur's eldest sister staying here. After luncheon the Bishop took us out driving in the beautiful country, to Shepton Mallet, going through the pretty and picturesque villages of Dulcot and Croscombe, and returning by another road over higher ground. A very pretty drive, with beautiful views of the distance to S. and W., as far as the Quantock Hills and Brean Down.

The Bishop's Palace is indeed a noble old house, somewhat intermediate in character (as I mentioned last year), between a baronial castle and a mansion.

The Herveys have made great improvements in it since we were here last year, and the apartments are now very comfortable as well as handsome. The stately gallery, which extends along one side of the house, adorned with a long series of portraits of

1871. the Bishops, has been fitted up as a drawing room,
for which it answers extremely well. At one end of
it is a descent to the garden by an external stone
staircase, old and handsome. But the most
remarkable improvement has been the conversion of
the "crypt" (or more properly *undercroft*,—this I
understand is its technical name), into a really fine
and convenient dining hall. It had been entirely
neglected and filled up with rubbish : nothing could
appear less promising : and when Lord Arthur
formed the plan of converting it into a dining hall,
almost all his friends thought it a hopeless scheme.
It has, however, been perfectly successful. The
room is very long and low, with a groined roof, the
intersecting arches radiating from a range of low
thick pillars along the middle, and others along the
wall on each side : the light supplied by small and
narrow windows on one side only. Yet, so well has
it been fitted up, that it is not gloomy. It makes
really a comfortable room, as well as one of a very
peculiar, picturesque and striking character.

The garden of the palace is now in great beauty—
the brilliant green turf, the glowing flower beds, the
fine trees, and the noble and venerable buildings,
make an admirable combination. Besides the
exotic trees which I noted last year, there is a
superb Magnolia grandiflora, now covered with
blossoms against the walls of the house ; a Paulo-
winia, of considerable size, which I am told has
flowered ; and a very handsome Ilex.

Attended morning service in the Cathedral. In the afternoon Sarah and Kate showed us some of the apartments of the house ; and afterwards I had a delightful walk with those two girls across the " Park." and over the Tor hill. The evening was glorious and the effects of the sunset light and colouring indescribably beautiful.

August 28th.

We had a delightful little expedition, with the two girls, their brothers Constantine and Jemmy, and Lady Augusta, to a pretty little green valley on the slope of the Mendips, below the village of West Horrington, between the Bath and Bristol roads. We sat down on the grass and drank tea in most rural style. Spiranthes autumnalis, the Lady's Tresses, was abundantly in flower on these grassy slopes— more abundant and of larger size than I have seen it elsewhere.

August 29th.

We drove with Sarah to Glastonbury, to meet the Bishop who had gone the day before to Yeovil, to consecrate a Church.

We sat some time amidst the beautiful ruins of the Abbey.

August 30th.

The Bishop went away again early to a distant part of his diocese.

I had a very agreeable walk with the girls to

1871. Milton Hill, a mile or so north-west of Wells, in the
direction of Wookey; it is a rounded green hill, an
outlier of the Mendips, partly wooded but chiefly in
pasture; and its top commands an extensive and
fine view. The distance was hazy, but we could
make out Brean Down, and even the Steep Holm.

August 31st.

I went out with Sarah, Lady Augusta, and the
Bishop (Fanny had hurt herself a little by a fall in
the garden, and found it prudent to stay at home).
We drove past the paper mill at Wookey, then
ascended a long, steep, stony lane, and came out
on a pleasant, open hill-pasture, opposite, and fully
in view of the fine, rocky ravine of Ebber. Here we
strolled about, enjoyed the extensive view over the
beautiful country, and drank tea. The afternoon
was fine, and very picturesque, the clouds beautifully varied in their forms and grouping, casting an
admirable variety of shadows; the sunshine
streaming down through the openings and bringing
out in brilliant relief, now one part and now another
of the landscape. It is a delightful country, and
was enjoyed in delightful company.

Sarah pointed out to me the Duke of Wellington's
pillar on Blackdown, Quantock's Head, Dunkery
Beacon, and Bossington Point, beyond Minehead;
but we could not make out the Foreland near
Lynton, which is sometimes visible from this spot.

September 1st.

We both went with the Bishop and Lady

Arthur, Lady Augusta, Sarah and Kate, and 1871. Constantine, to a spot on the brow of the Mendips, above the village of Easton, where we drank tea on the grass as we had done the day before.   But though the morning had been splendid, the weather clouded over, and a chill wind with threatenings of rain drove us from our station.   The ascent to it had been bad for the horses and carriages, and for those in the carriages ; but the descent was worse.

———

September 2nd.

Our delightful—most delightful—visit at Wells, came to an end, and we parted sorrowfully, from our dear friends, the Herveys.   It would hardly be possible to enjoy a week more than I have that which I have passed here.

Railway journey from Wells by Glastonbury (where we changed trains, and got on the Somerset and Dorset line), Wincanton, Temple Combe, Sturminster, Newton, Blandford, Wimborne Minster, to Poole.   There flys which we had ordered were in readiness, and we had a pleasant drive of about 4¾ miles to Bournemouth, over a country glowing with heath and furze in full blossom.

———

September 3rd to 7th.

At Bournemouth.   Not very interesting as far as the shore is concerned.   Cliffs of mere sand, moderately high, extend for several miles on each side of the opening, which is properly Bourne-

1871. mouth ; they are strangely torn and furrowed, and cut into strange shapes by the action of water.

The sand generally yellow of various shades ; some beds pure white and very fine, like that of Alum Bay. In some places a dark grey clay, much encrusted with sulphur from the decomposition of pyrites.

I saw also, in some places in the cliff, abundance of round, smooth, well rolled white coated flint pebbles, such as Lyell describes as characteristic of the Woolwich beds.

I saw no maritime plants, except Psamma arenaria and nothing cast up by the sea except Zostera.

Good views of the Isle of Wight to left, and Purbeck (nearer) to right. The high chalk cliffs of the Needle Point, shining brightly in the sunshine, make a fine appearance on a clear day, and beyond them we see distinctly the high downs towards the centre of the island, and the outline of the south-western coast to St. Catharine's Point. Inland, especially towards Poole, the country is pretty, fine, open, hilly heaths, beautiful at this season from the profuse blossoms of the heath and furze, and diversified by extensive plantations of Scotch Pine and Pinaster.

On the 5th we drove seven miles to call on Lady Sydney Osborne at their cottage called " The Hive," situated on the very shore of Poole harbour, near its mouth. We did not see her, but enjoyed a pleasant drive, delightful air and most agreeable views over the varied, hilly country, and over the

expanse of Poole harbour, with its branches, inlets 1871.
and islands.

On the 4th we drove (in rain) to Christchurch,
eight miles distant, and saw the magnificent
Church, which would be well worthy to be a
cathedral.

The Norman nave: the very lofty choir with its
beautiful roof, the old and curious *reredos*, with its
rich and elaborate sculpture representing Jesse,
David and Solomon, the Adoration of the Magi,
&c.; the beautiful little sepulchral chapel of the
unfortunate Margaret, Countess of Salisbury :—all
these are very interesting. The monument to
Shelley and his wife is fine.

September 7th.

From Bournemouth to Oaklands.

September 8th to 11th.

A pleasant, quiet time at Oaklands with dear
Minnie and Sarah, who are like a sister and niece to
us, and good kind Lady Napier. On the 8th
Minnie took us to see Leigh Park, formerly Sir
George Staunton's, now belonging to Mr. Stone,
M. P. for Portsmouth. Mrs. Stone, who is very
young and pretty, is a daughter of Mr. Helps,
(*Friends in Council*).

LETTER.

1871.   My Dear Henry,

Many thanks for your letter of the 31st.   I am very glad to hear by Cissy's letter, which Fanny received this morning, that you are the better for your visit to the Highlands, as I hoped you would be.   It is very unlucky that we should just miss seeing dear Harry during his short visit to England, and I am very sorry for it, but I hope we shall certainly see him at Barton in November.   It must have been very pleasant having him with you in Scotland, and it is delightful to hear such favourable accounts of him.   I fully trust he will grow up everything that you and we can wish.

It would be difficult to enjoy anything more than I did our week at Wells,—it was perfectly delightful.

The Arthur Herveys have made the Palace into a really comfortable as well as very handsome and stately house, and the garden is very beautiful. Here we are spending a few days happily.

I have a letter from Edward, from *Pesth* in Hungary, of the 3rd: he seems to have got on very successfully so far.   He promises to write from Odessa.   He says Pesth is so much *improved* (since he was there in '33), that it has lost almost all that rendered it interesting to a stranger.

Fanny is better since we came here.   I am perfectly well, only a little stiff in the joints, as it is

not unnatural in my 63rd year.   Not that anything 1871.
like it is the case with Lord Arthur, who is as active
as a young man.

With much love to dear Cissy and Emmie,
believe me,

Ever your affectionate brother,
CHARLES J. F. BUNBURY.

JOURNAL.

September 11th.

Up to London in company with Minnie and
Sarah.

September 12th.

Emily Napier, Clement and Herbert dined with
us.

September 14th.

Down to Barton with Clement:—arrived safe and
well at our dear home—thank God.

LETTER.

Barton,
September 15th, 1871.

My dear Joanna,

I have had great pleasure in hearing of
your mode of life and occupations at Innsbruck, and
the pleasant time that you have been passing with

1871. the Pertzes, and I am very glad that they are going to pay you a visit at Florence. It is very long since I saw Innsbruck, and then only for two or three days, but I have a general impression of a very picturesque town, and very striking scenery.

Poor Sir Roderick Murchison, I fear, is going ; Fanny enquired at his door when we were in town the other day, and heard that he was much worse lately, and had lost his speech. I fear he will not live to welcome Dr. Livingstone back to England. A good and valuable man is lost in him.

It is a great comfort to hear that Charles Lyell is so much the better for their tour ; I did not feel at all easy about him when we were in London in the summer.

I have not read a great deal since May, the principal things have been :—Kingsley's " Christmas in the West Indies," which is a very pretty book, perhaps a little diffuse, and with rather too much botany in it for some reading, (not at all too much for me, as you may suppose), but both entertaining and instructive, and telling much about the West Indies, which I at least did not know before. Francis Galton on " Hereditary Genius,"—not con- clusive, I think, but very curious and ingenious, and showing immense industry. Sir Thomas Wyse's " Impressions of Greece,"—very agreeable reading.

Nassau Senior's " Journals in France and Italy, 1848-52," very interesting, but certainly one must take his reports with certain allowances, as Susan remarked in a letter to Fanny. You see my reading appears rather desultory. I am now engaged with

the last edition of Milman's "History of the Jews." 1871.
I have not yet made much progress with Darwin's
new book, but hope to set to work at it in earnest
when the long evenings begin. I shall then draw
up my scheme of reading for the winter, and try to
keep to it.

I have not seen Dr. Hooker since his return from
Morocco, but have read some very interesting
notices of his expedition, in the *Gardeners' Chronicle*
and in the *Geographical Society's Proceedings*. He
has made a fine botanical collection, and ascertained several new and curious facts relating to the
climate, &c. of the Atlas. It is very fortunate that
he has returned safe, for the government and the
manners of the people are horribly barbarous.
What a lull in political affairs after the terrible
excitement of the last two years! I wonder
whether France will settle down peaceably under its
new Government—and for how long? I cannot
say that I have much confidence in the new
President of the Republic (Thiers); his past career
does not inspire me with much, and I believe that his
"Histoire du Consulat et de l'Empire" had a great
share in cherishing that ambitious and encroaching
spirit which led to the last war with all its
disasters. One of the great difficulties of France,
which has become very conspicuous in the last year,
is—that the French really form two distinct nations:
the peasantry, who have the religious beliefs of the
dark ages, with education and enlightenment to
match: and the people of the great cities, who have
an excess of mental activity, with a constant craving

1871. for change and excitement, with no belief in God or man.

Give my love to dear Susan, also to Leonora and her husband and children. Annie and Dora must be very interesting girls, and I should much like to see them.

Believe me ever,

Your very affectionate brother,

CHARLES J. F. BUNBURY.

September 18th.

***

JOURNAL.

September 20th.

Fanny received in letters, from both Sarah and Kate Hervey, specimens of Gentiana Pneumonanthe, gathered near Bournemouth or Poole, by Lady Augusta Seymour. They are very small, like all the English specimens of that Gentian which I have seen ; small, I mean, in comparison with those I have gathered in Switzerland. I have never, myself, found it growing in England.

***

September 22nd,

The newspapers of last Monday contained an account of the ceremonious opening of the Mont Cenis Tunnel. The successful completion of this work is certainly a great triumph of engineering skill and science. I would, however, rather not pass through it. The Tunnel was begun on the 15th of August, 1857, and finished on the 26th of December, 1870.

Received a letter from Edward from Odessa. He 1871.
says :—

"I was even more struck than before with the
"scenery on the narrow pass of the Danube, from
"Moldova to Orsova, which is certainly the finest
"that I know on any river. The rest of the course
"of the river is comparatively tame and uninteresting,
"but still it is by no means without interest to any
"one who has eyes to see, and a disposition to
"use them. There were several picturesque
"Turkish towns, especially Nicopoli, and very
"picturesque figures to be seen at the landing places
"where the steamer touched. I was also much
"pleased with seeing some *pelicans*, which I had
"never seen wild before ; in one place there were
"more than a hundred of them assembled on a
"sand bank in the river. Even the lowest part
"of the Danube, where it turns north to Galatz and
"thence to the sea, is not so utterly flat and marshy
"as I had expected, and I was surprised to find
"that the Dobrudscha (so famous, or rather infamous,
"for its unhealthiness), contains a considerable range
"of hills, rising almost into mountains, which extend
"through its whole length."

LETTER.

Barton,

September 23rd, 1871.

My dear Edward,

I will begin a letter to you to be finished when
1871. I hear from yon *whither* it is to be sent.   I have received both your letters from Pesth, Sept. 3, and
from Odessa, Sept. 10, and thank you very much
for them.   I was much interested by your observations on the bank of the Danube.   It must have
been very interesting to see pelicans in their wild
state ; it is (as I know by experience), a great
pleasure to see for the first time animals alive
and wild, which one has been used to see in
museums or menageries.

I am very glad indeed that your journey hitherto
has been so prosperous, and I hope it will continue
to be so.   We were away from home, as we had
planned, exactly three weeks from Thursday,
August 24th to Thursday, September 14th.   We
spent a most delightful week at Wells—as enjoyable
as it could possibly be.   Lady Augusta Seymour
was staying with the Arthur Herveys, and we got
much more acquainted with her than we had ever
been before, and liked her very much ; Mr. and Mrs.
Skrine were also there for two days, and were very
pleasant — *she* was one of the Millses of Saxham,
Susan Mills, I think.   Thence we went to Bournemouth.

Bournemouth disappointed me on the whole, 1871. though we had a very pretty and agreeable drive over the heaths to call on Lady Sidney Osborne, who lives near the mouth of Poole Harbour; and we were also much interested by the fine old Church at Christchurch. Then we went to Oaklands, spent four days very pleasantly with Minnie and Sarah; up with them to London (they were going to Miss Strutt's wedding), and so home.

Believe me ever your very affectionate brother,

CHARLES J. F. BUNBURY.

JOURNAL.

September 28th.

Shocked at hearing of the early death of Mrs. Phillimore, the widow of John Phillimore. We have seen little of her of late years; nothing I think for the last two; but at one time we used to meet her often in London, and liked her much, and there was a great friendship between her and Susan Horner. She was a remarkably pleasing, interesting, attractive person, and I should think very amiable.

September 29th.

My Barton rent audit, very satisfactory. Henry has sent me a letter he has had from Edward from Sebastopol, whither he had gone from Odessa. In it he says,—" The first aspect of Sebastopol from the sea "disappointed me; the hills were lower and the

1871. " natural features altogether less marked than I
" expected. But its appearance when one lands is
" most striking and remarkable ; I had no idea that
" even at the time of the siege the town could have
" been reduced so completely to ruin ; far less that
" after the lapse of fifteen years it could still be, as
" it literally is, one mass of ruins. The guide-books
" compare its appearance to that of Pompeii, and
" the comparison is really not without truth, though
" it of course wants all the decorative features of
" Pompeii ; but nowhere else does one go through
" street after street of houses all in utter ruin. Then
" the love of the Russians for architectural display
" adds to the effect—porticoes and triumphal arches,
" and imitations of Greek temples, all standing but all
" knocked and chipped about and all looking as if
" they were ready to fall down at any moment."—
" The curious thing about its present aspect is that
" the ruins all look so *fresh*, one would suppose that
" the bombardment had been an affair of last year
" at farthest." " The line of intrenchments can
" still be traced, but it is very much less distinct and
" perfect than many Roman and British camps that
" I have seen in England, or than the Devil's Ditch
" at Newmarket. The Malakhoff is the best pre-
" served, the foundations of the old tower remaining,
" with its casemate quite perfect, and the Russians
" having a cemetery close by, there is a guardian
" appointed to prevent further dilapidations. The
" view from thence is the most interesting and com-
" plete that one obtains anywhere, and shows at
" once, the commanding importance of the position.

"I took a boat this morning up the harbour to 1871.
"Inkermann, and walked up from the bridge to the
"field of battle on the plateau above. It is now
"marked by a monument in honour of the slain of
"all three armies. Thence I made my way across
"the plateau to the Malakhoff, and had a very
"rough walk, fully learning by experience the truth
"of what I remember hearing from you and others at
"the time of the siege, of the difficult character of
"the ground, and the way it is cut up by ravines."

October 3rd.

We dined with Lady Hoste and Mr. Greene, at
Ixworth Abbey.

October 7th.

The Louis Mallets and Miss Pellew dined with
us.

Sunday, October 8th.

We went to morning Church.—Mr. Harry Jones
preached : his sermon excellent.

October 10th.

Fanny went to Mildenhall, and returned in good
time. I walked to the Shrub—found nothing.

LETTER.

My dear Lyell,

1871.

I was very glad to hear of your safe return home, and especially that you were so much the better for your tour ; I heartily hope the benefit will be permanent. That north-west country must be very interesting and agreeable, and I should like extremely to see it. Much of it is described in a very pleasant way in Gray's Letters. I hope in due time I shall hear something of your observations upon it.

I see in *Nature* there is a good deal of controversy going on about Professor Williamson's views of the structure of Lepidodendron, &c. I am not well *up* to the newest doctrines concerning the anatomy of plants, and am especially behind-hand as to the German writings; but, as far as I can judge, it appears to me that Williamson is in the wrong, and Carruthers and MacNab in the right. That is, I think there is no reason to consider the structure of the stems of Lepidodendron, Sigillaria and so forth, as truly Exogenous, in the sense in which that word is understood by botanists, as applying to the stems of ordinary trees. They may very possibly have increased in diameter, like those of Dracæna Draco and the arborescent Aloes, which, nevertheless, are essentially Endogenous. But still I am quite satisfied that I did right in recommending

that Williamson's papers should be printed, for they 1871.
contain a great mass of curious and careful and
minute observations, which ought not to be lost,
and it is of little consequence whether his theoretical
opinions are correct.

Fanny is pretty well and more easy in her mind
about Mrs. Stocks* who is comfortably settled at
Mildenhall, and seems to be better.

I am quite well ; weather very fine, but the nights
very cold.

With much love to dear Mary,

Believe me ever yours affectionately,

CHARLES J. F. BUNBURY.

JOURNAL.

October 13th.

Attended a special meeting of the Thingoe Board
of Guardians on a sanitary question.

We dined with the Wilsons. Met Lord and Lady
Middleton, Sir Richard Kindersley, Lady Cullum,
P. Blake, Henry Wilson.

Dear Minnie and Sarah arrived.

LETTER.

Barton, Bury St. Edmund's,
October 14th, '71.

My Dear Katharine,

I thank you much for your letter of the
day before yesterday. Since then you have heard

---

* The housekeeper, a very good woman.

1871. of the death of poor little Zoé.* I am very *very* sorry for the loss of the little creature; it was so tame, so familiar, so amiable, so caressing, so exceedingly attached to Fanny, it is no wonder we were fond of it; I never saw another bird at all like it in mental qualities. Fanny is very unhappy. I am glad however that it died beside her, not in her absence and that its death was so quick and painless.

I do not think I have written to you since I received your letter from the Isle of Arran (of Sept. 22). I hope you enjoyed your visit to that island, which I always think must be a remarkably interesting one. One of the thousand-and-one omissions I regret is, not having visited it. Did you find any plants there?

I have lately read a very able and important paper by Professor Geikie, on the igneous rocks of the Isle of Eigg. These very fine specimens of Asplenium marinum, which you gave me some years ago were collected (I remember) in the Isle of Eigg; and Dr. Somerville gave me a bit of the pitchstone from that island when I was fifteen.

Geikie shows conclusively that what is now a lofty abrupt ridge of rock cresting a great hill, was originally part of a lava stream running into a valley. What a transformation.

We have dear Minnie and Sarah here now, and hope to have a quiet, comfortable week with them before other company come.

The weather lately has been beautiful and most enjoyable: that clear, bright, still, yet bracing

* A Paroquet.

weather which we often have in October.—No frost 1871.
yet sufficient to kill the heliotropes, or to cut the
leaves of the castor-oil plant.   We have lately made
an improvement in the middle division of the old
greenhouse, by forming a bark bed in the centre,
which will allow, I hope, of the successful cultivation
of really tropical plants.   Of your Indian orchids, I
believe one or two are alive, but I am afraid not
flourishing.   I have a good dried specimen of Suæda
fruticosa for you : the plant takes a long time to
dry : I gathered it on the 5th of September, and it
was not thoroughly dried till after the 5th of
October.

Your account of your children and their studies
was very interesting and pleasant to me.

Believe me ever,

Your affectionate brother,

CHARLES J. F. BUNBURY.

JOURNAL.

October 17th.

Took a walk with Minnie.   We all went to hear
Lord John Hervey's lecture at the Bury Athenæum,
on his Travels in Palestine and America—good.

October 18th.

Fanny and Sarah went to Mildenhall, and
returned to dinner.   Had a pleasant walk with
Minnie.   The Louis Mallets dined with us.

Received a very interesting letter from Edward, from Kutais, on the river Rion, the ancient Phasis.

LETTER.

My dear Edward,

You will be sorry to hear of the death of Mrs. Phillimore, of which we received the news from the poor boy her son, on the 28th of September. I had seen her very seldom for many years past, but she was a remarkably pleasing, interesting, attractive person, and we were much shocked and grieved at hearing of her departure.

The early death of poor young Lady Sudley is likewise very saddening.

(October 20th.) Yesterday I received your extremely interesting letter from Kutais, for which I thank you very much ; but you do not tell me whither I can direct to you. I was a good deal relieved by this letter, for I had become rather anxious about you, none of us having heard for a good while before ; and we· had such storms just after the arrival of your letter from Sebastopol, that I feared you might have suffered much from bad weather on the notorious Euxine. I am very glad indeed that the Black Sea treated you so kindly, and I hope you will have equally good fortune in the rest of your travels. The voyage along the Circassian coast must have been very interesting, as well as the

scenery of the Phasis, and the picturesque inhabitants of the country.

Pray, when you reach the Caucasus, do not be too adventurous, but remember that "A neck that's once broken can never be set."

I have just been reading Tyndall's "Hours of Exercise in the Alps," and therefore my imagination is strongly impressed with the perils of Alpine adventure.

(October 26th.) Just received your extremely interesting letter from Tiflis, for which many thanks ; and in accordance with what you say in it, I shall dispatch this to Constantinople. I am delighted that you are getting on so prosperously on your travels, and so much pleased with Georgia.

We are here in the thick of our gaieties,—the house full of people, mostly young; and last night we had our dance; and I danced Sir Roger de Coverley with Lady Cullum : wherefore I feel rather *seedy* and lazy this morning. The Bury Ball is to be to-night, but I shall stay at home. Henry and Cissy and Emmy are with us, and Emmy seems to enjoy the gaieties exceedingly. Otherwise the party is mostly the same as last year, the chief new element in it being Mrs. Robert Anstruther, an extremely pretty bride—one of the Fitzroys : her husband is Colonel Anstruther's eldest son, an intelligent and scientific young officer. Henry seems much better in health than when he was in London in June. In the last week, both Babbage and Murchison have departed ; both born (it seems) in the same year, and dying within two or three days

1871. of the same time.   Murchison, old as he was, will be much missed, and will with difficulty be replaced, I should think, either at the Jermyn Street Museum or the Geographical.   But his death was not unexpected, for when we last passed through London, Fanny inquiring at his door, was told that he had had another attack and had lost his speech. There was therefore no hope of real recovery, and one could not wish his life to be prolonged in such a condition.   Lyell is returned in much better health, I am happy to hear, than when he left London. Fanny is tolerably well, only as usual apt to over-fatigue herself ; she sends you much love.

Believe me ever,
Your very affectionate brother,
CHARLES J. F. BUNBURY.

Much love, dear Edward, from me and from Sarah, who is with me.   We shall be so glad to see you again in England.   We miss you very much this party.—F. J. B.

JOURNAL.

October 21st.

Henry, Cissy, Emmy and Clement arrived.

Mary Lyell sends us the news of Mr. Babbage's death.   He was a very old man, and had been for a good while in bad health, so that there was nothing unexpected in his death.   He was a man of un-questionable genius, but a very odd one.

Archdeacon Chapman dined with us. Mr. and Mrs. Mills, Ladies Octavia and Wilhelmina Legge, Minnie Powys, Mr. and Mrs. John Paley and John Hervey arrived.

News of the death of Murchison. I do not think he was equal in mental power to Lyell; he had not the same combination of large and comprehensive general views with true scientific caution. But he had a very considerable degree of systematizing genius; he had great powers of observation, he had indefatigable activity and perseverance, and by these qualities he was enabled to render pre-eminent services to the sciences of geology and geography. In fact I should say that hardly any scientific man of his time was more useful. His Silurian System, and "Siluria," his work on Russia, his paper (or rather work) on the Alps, and many of his addresses as President of the Geographical Society, will preserve his name in perpetual remembrance. Perhaps I have, in what I said before, rather underrated his genius :—his anticipation of the Australian gold, and of the geographical formation of the interior of Africa, were scientific achievements of a high order.

———— ————

October 24th.

Attended the Quarter Sessions at Bury—question about a new Petty Sessional Division—long debate.

————

October 25th.

Walked with Mr. Mills round the arboretum, &c.

1871.   Mr. and Mrs. R. Anstruther arrived.   Our dance —very successful—a large number of guests—much animation and appearance of enjoyment.

October 26th.

Received a most interesting letter from Edward, from Tiflis.

All the party, except Mr. and Mrs. Mills and myself, went to the Bury Ball.

October 27th.

Walked about the grounds with Mr. and Mrs. Mills.   General and Mrs. Ellice dined with us.   A little dance in the library.

October 28th.

Walked with Mr. Mills.   Mr. Praed, Mr. Burrell, and Captain Grant, went away.   Lord Augustus Hervey, Admiral Spencer, and George Napier arrived.

October 30th.

The house full, for the last week, with a very gay and pleasant party : in the main, nearly the same party as at the same time last year.

October 31st.

We attended the Church Conference under the presidency of the Bishop of Ely,* at Bury.   It was

* Harold Browne.

well attended. The Bishop's address was very 1871. good; characterized by the same merits which I noted in '68.

Mr. Rodwell spoke very well; and Mr. Percy Smith. I spoke, but hastily, without due preparation and not quite to my satisfaction, though I received several compliments upon it.

The weather still very "open" — no real frost hitherto, so that the Dahlias and other half hardy plants are still unhurt in the open beds, and the leaves of the Castor-oil plants (Ricini) which have been planted out, show no signs of injury by frost; nor do even the blossoms of the Fuchsias, in situations a little more sheltered. The leaves of the trees have fallen very fast within the last week or ten days, and many are quite stripped, though many are still well clothed, and some even green.

November 1st.

Fanny went with the Millses and others to see Hengrave. Lady Mary and Lord John Hervey arrived; Also Mr. Burrell.

November 2nd.

Walked about the grounds with the young ladies, and the two John Herveys—great merriment in the arboretum, with hide-and-seek, &c.

All the young party, with Minnie and Admiral Spencer went off the Thetford ball.

November 3rd.

Lady Mary and Lord John Hervey went away.
Lady Gage and her party came to luncheon.

LETTER.

Barton, Bury St. Edmund's,
6th November, 1871.

My dear Katharine,

Our delightful party broke up on Saturday,
and we have a little interval in our gaieties ; but
only a very short one, for another party are coming
to-morrow—a more learned party.   Henry, Cecilia
and Emily and George Napier are here still.   Our
fortnight of gaieties was extremely agreeable, and
everything turned out as well as possible ; our dear
old Mr. and Mrs. Mills (he in his 80th year) were as
charming as ever.

I think Geikie's " Memoir of Murchison," in
Nature very well done.   Sir Roderick will be much
missed, though his former attack had prepared us
for his departure, and loosened his ties to society.
It would not perhaps be quite strictly correct to call
him a great man, but he was a most eminently
useful one ; few men of science of his day perhaps
more so ; and his kindly and conciliatory disposition
must have contributed much to his usefulness. Both
at the Jermyn Street institution, and at the Geographical, he was, I take it, emphatically the right
man in the right place.   And in the records both of
geology and of scientific geography, his name will

always stand very high. This is certainly a remark- 1871.
able year, in which Herschel, Babbage and
Murchison have all taken their departure.

As soon as we are alone again, I mean to set to
at Jowett's "Plato," which I have been incited to
read by an article in the last *Edinburgh*. In the
mean time I am amusing myself with Sir James
Stephen's "Ecclesiastical Biographies," which are
very agreeable reading, and with Tyndall's "Hours
of Exercise in the Alps."

We had beautiful weather, all the latter part of
October, and enjoyed it much; it was so mild that
all "the bedded out" plants remained unhurt in the
beds till yesterday; but now at last a decided frost
has come to demolish them all. The conservatory is
full of beautiful flowers, but nothing particularly new.

Fanny is pretty well, but excessively busy,
working hard at arrangements for Arthur's education
as well as at the social ones.

It is always a pleasure to hear from you, and I am
very glad to have such a comfortable account.

Believe me ever,

Your very affectionate brother,

CHARLES J. F. BUNBURY.

JOURNAL.

November 9th.

Dear Henry and Cissy and Emmy and George
Napier went away.

During this week (from the 7th), we have again had our house filled with a very agreeable party, of a more grave and learned character than the former one :—Sir Bartle Frere and Lady and Miss Frere, Lord Charles Hervey and his daughters Susan and Isabel, Lady Rayleigh, John Strutt and his bride, Mr. Twistleton, and (latterly) William Napier, also Mr. Lott and Mr. Grant.

Sir Bartle Frere is a delightful man. I liked him from the time I first met him, and my opinion of his agreeable and estimable qualities is raised in proportion as I know him better. With vast knowledge (especially of geography and of everything relating to the Eastern world), deep reverence for religion, a great and noble character (as has been amply shown in his public life), he has a peculiar softness and gentleness of manner, a courtesy, a modesty, which are peculiarly pleasing in so eminent a man. He talks readily, and his conversation is full of matter.

Of Lord Charles Hervey I spoke in my Journal of last February, and my opinion of him is not lowered. His daughters are well informed and pleasant girls. Mr. Twisleton is a great talker, full of anecdote, and extremely entertaining. After the ladies had gone upstairs, he kept a party of us men listening for a long time to his amusing anecdotes of Lord Brougham, Lord Glenelg, Lord Palmerston and others, which he told admirably well.

Sir Bartle Frere was much gratified during his stay here, by receiving official intimation that the

Government have granted a pension of £300 to Dr. 1871.
Livingstone's family. Livingstone is a man so
disinterested, that he has never thought of doing
anything for his own family; and one of his
daughters is a cripple. Sir Bartle characterized
Livingstone as the *truest* man he had ever known.
I mentioned that my wife and I had heard Livingstone give a lecture at Bath (in '64), and had been
much impressed by his unmistakeable simplicity
and singleness of purpose; Sir Bartle said, those
are indeed qualities for which he is most remarkable. He believes that Livingstone is alive:—that
if he had perished we should certainly have received
distinct information of it through the natives. He
thinks that he is most probably working round by
the *west* of the great lakes (the two Nyanzas), and
that we shall hear of him first through Sir Samuel
Baker.

Sir Bartle Frere spoke of De Foe's " Captain
Singleton," (which I remember reading when a boy),
and said that De Foe's description of the interior of
Africa (in Singleton's imaginary journey across it),
shows him to have had general ideas approaching
on the whole, pretty nearly to the truth, as to the
geography of that continent. He thinks that De
Foe had probably known some naval or mercantile
adventurers, who might have picked up, from the
Portuguese on the African coast, some correct
notions on this subject. He said there is no doubt
that the Portuguese, long ago, had much more
correct general notions as to the geography of the
interior of Africa, than were received among

1871. professed geographer until very lately.   Central
Asia, also, he says, was certainly much more
accessible to Europeans in the middle ages than it
has been since.

He thinks highly of Col. Yule's edition of Marco
Polo.   Sir Bartle Frere thinks that Murchison can
hardly be replaced, — that his conciliatory and
pacifying influence will be sorely missed.   I mentioned Murchison's last Geographical Address as a
remarkable work under his circumstances : Sir
Bartle said that the Secretary, Mr. Clements
Markham, had been very useful to him ;—I understood him to imply, that Markham had had
an important share in the preparation of the
Address.

I talked with Mr. Twisleton about Junius.   He
said that the earlier inquirers into the authorship
had not attended to the letters published under
various *other* signatures by the same writer: perhaps
had not known of them any more than of his private
letters to George Grenville.   Junius, Mr. Twisleton
says, was certainly (at least for some time) an
admirer and partisan of Grenville,—and to such a
degree, that he strennously advocated the taxing of
the Americans, though he afterwards adopted the
views of Chatham.   This is alone sufficient to prove
that Burke was not Junius, as Johnson was inclined
to believe.

Though Mr. Twisleton has bestowed so much
time and thought and cost (for his book must have
cost him a great deal of money) on the Junius
question, he readily agreed with me that those

famous letters speak very ill for the moral character 1871.
and disposition of their writer.

A sharp frost.   Very cold, bright and clear.

William Napier left us.   He made an interesting
tour this autumn (immediately after the "Campaign,") visiting the localities of all the principal
battles of August and September, 1870—beginning
with Sedan and proceeding from thence by Metz,
Woerth, &c. to Strasbourg.   He says, however (and
so it strikes me) that it would have been better to
take them in the reverse order—the order in which
the battles were fought, beginning with Saarbruck,
and ending with Sedan.   He saw the field of Sedan
very well ;   the commandant (a Saxon) was very
polite to him, and not only lent him a horse, but
accompanied him over the field, and pointed out to
him, all the interesting spots.   This commandant
told him how it was that MacMahon was wounded
in that battle.   He, the commandant, having the
command (I think) of a brigade in the battle, and
having, according to his orders, taken up a certain
position early in the action—saw at some distance
in front of him, a body of troops whom he could not
well make out ;   he did not feel sure whether they
were French or Germans.   He therefore directed
the officer commanding the artillery of his brigade to
aim a shell, *not* into this mass of troops, but at a
tree very near them, so as to draw from them some
demonstration.   The artillery officer did so ;

1871. MacMahon and and his staff happened to be beside the tree, and the shell fell in the midst of them, killed two of the aide-de-camps, and dangerously wounding the Marshal himself.

William Napier says that the position of the French at Gravelotte (the most sanguinary of the battles) was very strong; in particular, their left was defended by a deep wooded ravine, and the Germans never succeeded in forcing that part of the position.

They gained the battle at last by overpowering and turning the French right; though there also the action was very severe. The German force which attacked the village of St. Privat at the extremity of the French right, lost 6,000 men in ten minutes.

Bazaine's error was (William says) that he was too anxious about his left, and kept the whole of the Imperial Guard, throughout the action, in reserve behind that part of his line, whereas, if he had carried it at the critical time to the support of his right, he might have baffled all the efforts of the enemy. At Woerth also, he says, the position of the French was very strong: but there they were very much outnumbered.

William Napier said that at the "Autumn Manœuvres" or "Aldershot Campaign" the General who did the best was Major General Lysons. The Duke of Cambridge (William says) knows how to *handle* troops, to manœuvre, better than any other man in the kingdom; he understands his profession thoroughly: but whether he has the nerve and coolness and presence of mind, necessary

for command in a great battle, may be more 1871.
doubtful:—he *had it not* in the Crimea.

William says there is no doubt that the French
were very inferior to their enemies in generalship
and in discipline : and these were the real and
simple causes of their defeats. This is exactly what
I had supposed.

William has come back from France with a
decided impression that there is still a strong
Buonapartist feeling, especially in the army and the
peasantry. Strange as it seems to us, they believe
that the Emperor was not to blame for the disasters
of the war (whether he was to blame for beginning
the war at all, they probably do not consider), and
they think also, that so long as he was in power, he
kept down the Communists.

November 14th.

We dined with the Benyons at Culford : met Mr.
Cox and Lady Wood, the Hortons, Bains and
others.

Friday 17th.

We dined with the Maitland Wilsons: met Col.
and Mrs. Anstruther, Miss Rowley, Lord John
Hervey, Mrs. Dawson and others.

Tuesday 21st.

Walked with Fanny and Scott to see Mr. Lofts'
new farm buildings.

A A

Monday 27th.

Our dear Kate Hervey is engaged to be married
to Mr. Charles Hoare, a marriage which seems to
give great satisfaction to her father and mother, and
to her sister Sarah. I most earnestly hope and
trust it will be a source of great and lasting
happiness to her ; I do not know any one more
deserving of happiness, or more qualified to make
others happy.

———

Tuesday 28th.

The Dean of Ely, and Mrs. and Miss Merivale
arrived. Mr. and Mrs. Percy Smith, Mr. and
Mrs. Victor Paley, and Miss Hoare dined with us.

———

Wednesday 29th.

The Merivales went to see Bury. Lady Cullum,
the Hortons, Mr. Abraham, Mr. B. Bevan and his
daughter dined with us.

———

Thursday 30th.

The weather this month has for the most part
been very severe. The first real frost of the season
was on the 6th, killing off the " bedded-out " plants,
but leaving the lemon-scented verbena and the
fuchsias unhurt. These were cut down by sharp
frost on the 12th and 13th. On the 17th came
violent snow storms—real snow, with excessive
cold ; several days following were bright and clear,
with hard frost, and in the night of the 18th-19th
the thermometer was 10 deg. below the freezing

point. On the 20th our ice house was filled, well 1871. and satisfactorily filled : very unusual in November. On the 23rd a thick cold fog, the 24th a cold thaw, and to-day again there have been repeated falls of snow.

I cannot help being sorry that the Government at Versailles have put Rossel to death. No doubt he was, according to strict military law, deserving of death, because, having been in the service of the Government, he had gone over to the insurgents, and fought against the Government which he had originally served. But in this he did nothing more than General Lee, and several other officers in the American civil war : and he seems to have been a fine high-minded man, not concerned in any of the murders or other atrocities of the Commune. What makes it more cruel is the executing him in company with a vile murderer like Ferré.

Friday, December 1st.

Very cold, a bitter wind. Went to the Board of Guardians of Thingoe Union—special meeting for election of Workhouse Master.

Saturday 2nd.

All our guests departed. We went (posting in our own carriage) to Ipswich to a meeting of Governors of Albert College, to elect a Head-Master:—Lord Stradbroke, Sir Edward Kerrison, Lord Bristol and many more—meeting satisfactorily conducted—Mr. Bird elected.

Since the 28th of last month we have had another
party of guests here : the Dean of Ely and Mrs. and
Miss Merivale, Mr. and Mrs. and Miss Gambier
Parry, Mr. Eddis (the artist), and Mrs. Dawson.
Mr. Gambier Parry is most interesting. He is a
lively, animated, eager man, of pleasant manners
and of very fluent and agreeable conversation :
decidedly an agreeable guest in a house ; a
perfect enthusiast in everything relating to the
fine arts, and highly accomplished in them. His
own paintings, with which he has decorated his
church at Highnam, near Gloucester, are (as Sarah
Hervey tells me in one of her letters), of extraordinary beauty.

Mr. Gambier Parry told this story of Lord
Brougham's readiness :—

Late in his life, Brougham was present at a
wedding breakfast, where, through some mis-
arrangement, he was unexpectedly requested to
return thanks for the bridesmaids. Not discon-
certed, he rose, and began :—" Ladies and gentlemen, whether it be on account of my youth, my
innocence, or my beauty," — this exordium threw
the whole company into such fits of laughter, that it
was unnecessary for him to say a word more.

LETTERS.

My Dear Joanna,

Very many thanks for your very agreeable letter of the 21st November, which gave me great pleasure. I leave it to Fanny to tell you all the particulars of our gaieties and sociabilities, only saying that I very much enjoyed both our ball party, and the later one of the Freres, Herveys and Strutts.

I was much interested by your account of the Borgia family, and the portrait of Cesare. As far as I have read Italian history, Cesare Borgia never appeared to me to be quite so bad as some of the Visconti; to be sure he might fall short of them, and yet be worse than most men. That family (Borgia) you know, produced an eminent saint as well as an eminent sinner. San Francesco Borgia; of whom I lately read in Sir James Stephen's "Ecclesiastical Biographies." I have lately read the greatest part of that book, not for the first time, for I read it several years ago; but I read it again with very great pleasure. It is to me a delightful book, the author is peculiarly successful, I think, in giving vivid and distinct ideas of the men of whom he treats, and his tone and spirit are extremely pleasing. The essays, which are my favourites, are those on "The French Benedictines," on "Wilber-

1871. force," and on " The Clapham Sect." Those on
"Hildebrand," and on " The Port Royalists" are
extremely interesting.   I have lately read with
great pleasure Tyndall's " Hours of Exercise in the
Alps." It is very interesting and particularly well
written; he describes the glaciers, peaks, &c., quite
beautifully, and his narrations of adventures and
".hair-breadth scapes" are most interesting.   I have
not yet finished Darwin, but am in the middle of the
second volume.   I agree with you in thinking that
he writes very well, and the candour and modesty
which he shows throughout are very agreeable.
That part of the book which relates directly to man,
appears to me very inconclusive; but the chapters
on what he calls " Sexual Selection " (much the
larger part of the book by the way), are very curious
and interesting and admirably well worked up.   After
all it might be more respectable to be descended
from a well-behaved monkey, than from Cesare
Borgia or Bernabo Visconti, or from some ruffianly
Captain of *Condottieri*.   Leigh Hunt says (and I
think he is right) that for a man " to *make a beast* of
himself " might sometimes be a decided rise in the
scale of creation.

Edward is on his way back from Georgia (the
Asiatic, not the American Georgia), where he has
seen a great deal that is interesting and picturesque.
I heard from him the other day from Marseilles,
where he was in quarantine (having come from Constantinople), and I hope he will be in England next
week.   He was delighted with the Circassian
coasts of the Black Sea, and with the beautiful

valley of the Rion (the ancient Phasis), about 1871.
Kutais ; and with the picturesqueness of the people.
He describes Tiflis and the population as the most
varied and picturesque   He was there at the same
time with the Emperor of Russia, and saw a great
fête, at which the Russian Court in all its splendour
was mingled with wild people from all parts of Asia,
Georgians, Persians, Tartars, Koords, &c. in their
various national costumes. We have just heard from
him from Paris, where he has arrived safe, after
being nearly frozen on the journey from Lyons, and
where he means to stay till the weather becomes
milder.

The cold at Paris according to the newspapers is
more intense than has been since 1789. Here too,
the weather is intensely cold, and the ground deep
in snow. We shall have an awful winter if it goes
on long as it has begun.

As I shall probably not write again to you this
year, I will now wish you and dear Susan (as I
do most earnestly and heartily) a very happy
Christmas and New Year, and many of them. I
was much interested by your account of the
Pertzes. Annie, in particular, must be a very
interesting girl. I hope we shall have a family
gathering here next year.

Pray give my cordial love to dear Susan, and
believe me ever your very affectionate brother,

CHARLES J. F. BUNBURY.

Barton, Bury St. Edmund's,

December 12th, 1871.

My dear Lyell,

Many thanks for your letter; it is a great pleasure to hear from you, and though you say nothing about yourself, I may hope that you have not lost any of the improved health which you gained by your autumn tour.

I am quite astonished to hear of the first volume of a new edition of your " Principles," already in the press. It seems but just the other day that I read your tenth edition, and it is quite fresh in my mind. The demand for another edition in so short a time is a very satisfactory proof of the usefulness and value of the work, but I hardly suppose you can have much new matter for this one. By the way though, I suppose you will draw largely upon Darwin's last book in your second volume. He has not convinced me as to Man (though he certainly shows that we have been in the habit of very unduly depreciating the Monkey family), but his chapters on " Sexual Selection " are very curious and interesting, and will always be valuable, I think, even if his main theory be set aside. But, to return to your letter, I did not know much about Dr. Carpenter's theory of currents, though I read his report (in the *R. S. Proceedings*) on the dredging voyage of the Porcupine, but I did not attend much to any but the natural history observations. I have also seen some discussions on the subject in *Nature*. It appears to me, as far as I can judge, a question to be settled by direct observation rather than by theory. I am glad that the result arrived at, is satisfactory to you. With respect to the further discoveries of fossil plants in the Arctic regions, I quite agree with what Joseph Hooker said to me last year—that those indubitable proofs which have of late years been discovered of the former existence of a rich vegetation in the Polar regions, constitute the greatest wonder and the greatest puzzle connected with natural history in the present stage of our knowledge. I shall like very much to see what you say about them.

I am sure you and Mary must be very anxious about the Prince of Wales, and feel deeply for his wife and mother. It seems almost wonderful that he is still alive ; yesterday and the day before we were almost every hour expecting the announcement that it was all over, but by the last telegram there seems to be again a slight gleam of hope. His illness seems to have awakened a very general good feeling in the country, and I hope will cause a real revival of the sentiment of loyalty. If he recovers, I trust he will become a more earnest and thoughtful man.

I hope neither you nor Mary have caught cold from the horrible cold weather we have lately had. I am tolerably well, which is as much as I could reasonably expect or hope ; and Fanny, I think, has stood the cold better than I have. I began to think we were falling back into the Glacial Period, and to apprehend that my house might be destroyed by a glacier : but happily the frost has broken up at

1871. last, and to-day has been comparatively mild and pleasant.

With much love from both of us to dear Mary and yourself, and wishing you both a merry Christmas and happy New Year, and many of them,

Believe me ever your's affectionately,

CHARLES J. F. BUNBURY.

JOURNAL.

Wednesday 13th.

Important talk with Scott on drainage of Conyer's Green cottages. Mrs. and Miss Ellice, Mr. and Mrs. and Miss Praed, Miss Bevan, John Hervey and Col. Tyrell arrived. Lady Cullum dined with us. The good news of Cecil's success at Sandhurst.

Friday 15th.

George Darwin and Charles Strutt arrived. Mrs. and Miss Horton and Mr. Abraham dined with us.

Saturday 16th.

For a week past we have been (like everybody else in this country), in great anxiety and alarm about the Prince of Wales. His illness began in November, but for some time he seemed to be going on favourably, and it was the relapse or increase of bad symptoms, early in December, which caused

such great and general and just alarm. In the 1871.
early part of this week his condition seemed almost
hopeless, and from hour to hour we expected the
telegram announcing his death. But he seems to
have rallied in a most surprising manner, and
though of course his state is still very dangerous,
there really appears some reasonable ground for
hope. His danger has had the effect of awakening
a most remarkable outburst of sympathy and loyal
feeling throughout the country. In all classes, and
almost without distinction of parties or creeds, the
manifestation of good feeling has been very striking
and very honourable to the nation. I trust that the
effect will be permanently good in the revival of
loyalty and the checking of those democratic
tendencies which have shown themselves of late.
And we may hope also, that if the Prince recovers,
it will be a turning point in his life, and he will
henceforth be a wiser and more earnest man.

——— ——

Monday. Dec. 18th.

The accounts of the Prince of Wales continue to
be very cheering.

Mary Lyell writes to us, that the general anxiety
about the Prince's condition, the strength of the
feeling and sympathy manifested by the lower
classes in London has been quite as great as the
newspapers represent it. She thinks this remarkable display of feeling is mainly owing to the
sympathy of the English with family attachments.

1871. All who have associated with the Prince, she says, have a liking for him, and she thinks it unreasonable to expect him to resemble his father, who was one in a million.

George Darwin, who has just left us, is lately returned from a visit to North America, in which he went by the Pacific railway to California. He told us that the Rocky Mountains, where that railway crosses them, do not form a chain of mountains in the ordinary sense; the slope up to the watershed is so gradual, that the ascent by the railway is almost insensible ; you are not aware that you are going up hill, and yet the highest point of the line is not less than 6,000 feet above the sea. The country is very bare and dreary, and there are no fine peaks or mountain ranges in sight. He described the Yosemité valley in California as being quite as extraordinary as I had heard before :— unlike anything he ever saw elsewhere. The sides of the valley are continuous precipices of 3,000 feet (and in some places much more), of vertical height. Looking from above, it seems as if you could throw a stone across the valley, though it is (I think he said), a mile-and-a-half broad. The upper parts of the mountains on each side, above the cliffs, appear as if rounded and smoothed by the action of ice, like the *roches moutonnées* of the Alps ; but in the valley he saw no signs of glacial operations.

Speaking of the mammoth trees (Sequoia gigantea) of California, he described their general form and appearance much in accordance with the sketch in

the *Bot. Mag.*, but said also (what was new to me), 1871.
that the bark of the trunk is of a bright orange-red
colour. He said that the enormous height and size
of the *generality* of the pines and firs in the
Californian forests were to him almost more striking
than the super-eminent size of the mammoth trees.
He thought that the average—the common run—of
those trees must be fully double in height and size
of our largest Scotch pines.

The State geologists of California (Darwin says)
are not of opinion that glacial action had any share in
the formation of the Yosemité valley. It is debated
among them whether it is to be ascribed entirely to
the action of water or (originally) to a great fracture
of the strata caused by subterranean movements.
Very much the same question which is so much
debated (with a more general application) among
our own geologists.

LETTER.

Barton, Bury St. Edmund's,
December 22nd, 1871.

My dear Katharine,

I am very glad to hear you have met
Charles Kingsley—how I wish I could see him. It
is close upon three years that we have been vainly
trying to induce them to come to Barton. In these
days of easy communication, it seems hard that

1871. people who have a regard for one another, and who live in the same island, can never meet.   And stupid people are so easily met.

I am very glad you thought Kingsley what I think him—fascinating.

Sir Bartle Frere is to my thinking another fascinating man.   I have, however, been rather disappointed with his Memoir of his uncle, Mr. Hookham Frere, which I have just been reading; not that it is not very well written, but the materials amount to less than I had expected. Mr. Hookham Frere seems altogether rather a disappointing sort of man; a very inferior man, I should think, to Sir Bartle.

I am going to read Mr. Frere's translations from Aristophanes, which I had often heard of but never seen before.   I am now close to the end of Darwin's book, and a very remarkable book it is.   I am not a convert to his main theory (at any rate as to more than the bodily frame), but it is really wonderful what a store of matter bearing on the question he has collected from all sorts of sources, and how admirably he has arranged it.   And his candour and perfect fairness, his modesty and love of truth, are very striking and admirable.   I must add, too, that the book has much increased my respect for monkeys.   How beautiful the illustrations are too.   I am reading Mr. Harness's Life with a good deal of pleasure ; there are many amusing anecdotes, and he seems to have been a wise and good man, and I do not like him the less for being rather old-fashioned.   I want to see the new " Life of Dickens," which seems to contain some extraordinary things.

I have been reading to Fanny, in the evenings, Miss Frere's "Deccan Legends," which are very prettily and gracefully written, and some of the stories amusing, and some curious.

We are very happy, as you may well suppose, about the prospects of the Prince of Wales' recovery, and greatly rejoiced at the general display of good feeling throughout the country.

I was very glad when I heard of Edward's safe arrival in London—safe from Constantinople and the cholera. He seems to have made a most interesting journey. Did you see him while he was in town? He is now gone to Lord Belper's, and we expect him here about the 12th of next month.

Joanna told me in her last letter that she had sent you a specimen of *Nothochlæna Marantæ* from (if I understand her rightly) the neighbourhood of Innsbruck;—if so, it is a more northern locality than any I had heard of: I did not know of its growing anywhere on the *north* side of the Alps.

I do not know whether I told you of the latest additions to my botanical library. 1, " Hortus Cliffortianus," one of the earliest works of Linnæus, and one of the scarcest, a fine book and a very good copy. 2, Lamarck and de Candolle's " Flore Française," the edition of 1815, in six volumes, a very good book, and rather a scarce one. 3, Harvey's " Phycologia Britannica."

1871.   I wish you and yours, with all my heart, a happy
Christmas and New Year, and very many of them.
With much love to your husband and Rosamond,
and your sons, and with every good wish,

Believe me ever,

Your very affectionate brother,

CHARLES J. F. BUNBURY.

December 23rd.

P. S.—We have sent two of our pictures (Mrs.
Bunbury and Mrs. Gwyn), to the Roy. Acad. Exhib.
of Old Masters.

JOURNAL.

Wednesday 27th.

Cecil, Clement and Herbert arrived.   A children's
party—the Christmas Pie : Mr. Smith assisted.

Saturday 30th.

We are very near to the end of the year, and
again I feel myself called to return most humble
and heart-felt thanks to Almighty God for His great
goodness in continuing to me, through this past
year, the many and inestimable blessings which I
enjoy.   I deeply feel that I can never be sufficiently
grateful.   The year 1871 has indeed been a very
happy one.   I had much anxiety for a time on
account of my wife's illness in London : but since
the middle of September, I am thankful to say, I
have been free from all uneasiness respecting her

health. With regard to all my causes for thankful- 1871. ness, I can only repeat what I wrote in my journal at the close of last year. We have enjoyed, during this year, a great deal of delightful society. Our circle of acquaintances has suffered some losses by death, though we are happy in having lost no intimate friends. The principal of those who have been taken away, are Sir John Herschel, Sir Roderick Murchison, Babbage, Thomas Lyell, Mrs. Phillimore, Lady William Hervey, Dean Alford. The first three of these were very eminent men, but they had lived to a ripe old age, and completed their life's work; Alford may be considered as having died prematurely. We have travelled little this year, and what new ideas I have gained have been through reading, conversation and correspondence, not through my own opportunities of observation. I have had several very valuable letters from Kingsley.

Sunday 31st.

Mr. Percy Smith read out from the pulpit the Queen's deeply interesting and touching letter (published in yesterday's papers), thanking the nation for the sympathy and good feeling they have shown during the illness of the Prince of Wales. It has indeed been very honourable to the people, and indicates, I trust, a renewal of the fine old sentiments of loyalty and attachment to the throne as the representative of Law and Order. The Queen's

1871. letter is admirably well expressed. It comes evidently straight from the heart, and goes straight to the heart.

<hr>

LETTER.

Extracts from a letter from Edward from Kutais.

September, 27th, 1871.

" I had heard much of the beauty of the south
" coast of the Crimea, and it has quite come up to
" my expectations.   The first part, crossing from
" Baidar to Alupka, is the boldest and grandest,
" with very fine, craggy mountains of limestone,
" presenting bold, rocky fronts to the sea, crowned
" by vast cliffs, with the rocky slopes descending
" from their foot to the water's edge, clothed with
" wild wood.   It is something like the undercliff of
" the Isle of Wight, on a vastly grander scale, the
" mountains rising to a height of 4,000 feet, and the
" precipices that crown them being quite worthy of
" the Alps.

" Alupka, where Prince Woronzow has a splendid
" villa, quite a palace in fact, is a lovely spot, and
" he has laid out very extensive gardens in the
" English style and planted with a great variety of
" exotic pines and shrubs of all kinds, magnolias,
" catalpas, &c.   But though the climate is very mild,
" from this part of the coast being so completely
" sheltered from the north winds; still it is nothing
" like that of the Riviera of Genoa.   Not only are
" there no orange trees or lemons, but even the

" olive will not flourish. Indeed I was surprised to 1871.
" see that the *indigenous* vegetation had very little of
" a southern character ; the wood that covered the
" slopes was almost all oak or horn-beam, mixed
" with maple, cornel, wild cherry, &c. The only
" thing that struck me as peculiar, was the extra-
" ordinary abundance and luxuriance of the wild
" vine, which often almost smothers the trees, and a
" beautiful creeper it is.

" All the way from Alupka to Yalka, and for some
" distance on the other side, is almost a succession
" of villas, including two imperial ones, and this
" tends to give to the scenery that life which it
" otherwise wants. The natural features are at
" least as fine as those of the Genoese Riviera—the
" mountains indeed a great deal finer—but one
" misses the picturesque towns and white houses
" and churches gleaming through the woods.

" I made one short excursion" (from Yalta) "inland
" towards the mountains, and saw a fine forest of
" Tauric pines in their original site ; but the axe of
" the woodman is already busily at work among
" them, and all around Yalta one sees the wild wood
" and forest that forms at present the great charm
" of the scenery, beginning rapidly to disappear.

" We left Yalta in the evening and arrived the
" next day at Kertch, where we stayed about 6 hours.
" Thence we steamed on along the Circassian coast
" to Sukhum Kalé and Poti, touching only at two
" small places called Nova Rossick and Taupse, the
" last a new place of about twenty houses, not to
" be found on any ordinary map, yet these two are

1871. "absolutely the only villages to be found along this
"whole line of coast, from Anapa to Sukhum Kalé,
"a distance of more than 200 miles.   Since the
"submission of the country to Russia, the Circassian
"inhabitants have emigrated *en masse* to the Turkish
"provinces; the Russians in consequence have
"abandoned their fortified posts, and there are now
"no inhabitants.

"It is a very rugged mountain country, but no
"very fine mountains and no considerable openings;
"and became extremely monotonous as we ad-
"vanced.   On the other hand, the voyage from
"Sukhum Kalé to Poti is most beautiful, with the
"glorious range of the snowy peaks of the Caucasus
"full in view all the time, rising above lower
"mountain masses all richly wooded, with the
"beautiful blue sea as the foreground.   We were
"certainly singularly fortunate, for the whole snowy
"range was visible without a cloud; even Mount
"Elburz itself, which lies far back, was seen to
"the greatest advantage in its vast mass of
"unbroken snow; it was about seventy miles
"distant.

"The Black Sea has hitherto been very kind
"to me, for I have not had an hour's sea-sickness
"since I have been on it. . . . . . This
"(Kutais) is a lovely place to be detained in : the
"hotel is just tolerable, but that one must put up
"with; and in all other respects the place is
"charming : in a beautiful situation on the river
"Riom (the ancient Phasis), here a fine brawling
"mountain stream, surrounded by green hills

" covered with partially cleared forest, with the most 1871.
" delightful winding foot-paths in all directions, and
" glorious views over the broad valley of the Phasis,
" and the mountains of Asia Minor beyond. These
" last are much finer, and approach much nearer
" than I had expected.

***

1872.

JOURNAL.

Tuesday, January 2nd.

The children's party and dance—a great merry- 1872.
making.

***

LETTER.

Barton, Bury St. Edmund's,
January 3rd, 1872.

My dear Katharine,

Many thanks for your very interesting
letter; all you tell me about Frank, and his
prospects, his peculiarities, and your views and
wishes for him, interests me extremely. I quite
enter into your anxieties and apprehensions. The
launching a young man in life, the choosing a profession or line of life for him, or helping him to
choose it for himself, is one of the most arduous and
anxious tasks that can fall to any one's lot. At the
same time, I feel sure that *your* son, with the blood
he has in him, educated with such loving care and

1872. watchfulness as I know he has been, with such
principles instilled into him, and with the examples
of his uncle and grandfather before him, *cannot* go
far wrong, morally, however far he may be away
from you in space. I see you are anxious because
Frank is of a different character and turn of mind
from your other children. But such differences are
in the course of nature, and there is room for very
great variations of that kind among good men. It
appears to me very natural that many young
English gentlemen should dislike a sedentary and
stay-at-home life, a life of desk work and other
monotonous drudgery, and should long for a more
free and adventurous life, for the freedom of a wild,
luxuriant, untamed country, and a glorious climate.
Tame as I am myself, I can enter in imagination
into the delights of such a career, and into a young
man's longing for it. Indeed I have long felt convinced, that with the great and still increasing
difficulty of providing in this country for all who
have been brought up as gentlemen, it would
become necessary for a large proportion of gentlemen's sons to go and seek their fortunes in distant
countries, as in the old Drake and Raleigh days.
Therefore I am not surprised at Frank's wishes,
and am glad that there seems to be such a fair
opening for him. What you tell me about Mr.
Smythies seems to hold out a good promise. Only
I hope Frank will be cautious, for it is difficult for
an Englishman to conceive how wild and lawless
and ungoverned the people of those countries are,
and how ready to use the knife. The climate of

that La Plata country is, I believe, delightful, and 1872.
there is every natural advantage—

"And only man is vile."

I do not know whether you are aware that
Maurice Kingsley began by trying that sort of life
some where in Uraguay, which is just the same sort
of country as La Plata : but he did not like it, and
afterwards went out to Colorado, in the United
States, where I believe he is getting on very well.
I daresay that Mrs. Kingsley, if you were to write
to her, would give you some information about her
son's experiences. Now I will stop. I hope you
will not think me very prosy, and still less that I
have wished to preach to you ; but I am very much
interested in what you tell me about your son, and
this has led me to expatiate. Whatever may be
decided, I hope and trust that the result may be
happy for you, and that Frank's future may be
everything you can wish.*

We are both well, and Fanny sends her love ; the
children's party yesterday was a great success.

Believe me ever,

Your very affectionate brother,

CHARLES J. F. BUNBURY.

JOURNAL.

Thursday, January 4th.

The servants' dance and merry-making.

Shocked by hearing of the sudden death of Sir
Edward Gage. He was a very pleasant neighbour,

---

* Frank was a very great favourite of ours.—(F. J. B.)

1872. a truly good-natured man, frank, kindly and genial, and will be much missed in the neighbourhood. I feel sincerely for poor Lady Gage, who was devoted to him.

———

January 5th.

News of the death of old Mr. Tyrell, the oldest gentlemen (I should suppose) in Suffolk, having almost completed his 96th year.

———

Sunday, January 7th.

Very cold. We went to morning Church and received the Sacrament. Yesterday between 1 and 2 p.m., there was a violent thunderstorm, accompanied by a most furious storm of hail,—one of the most violent I remember ever to have seen here. For a considerable time the ground was as white as if there had been a snowstorm, and to-day, 24 hours after, unmelted hailstones are still lying in heaps in shaded places. The thunder began (at least I first heard it), very soon after 1. At that time, and until the hailstorm broke over us, the blackness of the clouds and the lurid gloom over the country beneath them were almost awful.

———

Saturday, January 13th.

Edward arrived yesterday in good health and spirits.

———

The Bishop of Ely and Mrs. and Miss Browne, the Praeds and Mr. Lott arrived.

The Abrahams, Wilsons, Chapmans and Phillipses dined with us.

Edward tells me that there is a very good Museum of Natural History at Tiflis; and in this he saw specimens of the true royal tiger, and of the bison or aurochs, both killed in the forests of the Caucasus. The bison (identical with that of Lithuania), is a permanent resident in those forests. More than one species of wild goat or ibex also inhabits the Caucasus. The common pheasant, in a truly wild state, is common in that country, from which it was originally brought by Europeans; and Edward thought its flesh had a more *game* flavour than that of our semi-tame pheasants. He mentioned also a beautiful bird which he saw in the Museum, and which he was told was not uncommon in the forests of that country:—of the shape and character of a partridge, but larger than a common pheasant, and with beautifully spotted plumage. *Megaloperdix* was the name given to it in the Museum. Alfred Newton tells me that another generic name for this bird is Tetraogallus; that there is a Himalayan species, of which there are plenty of specimens in our Museums, but that the Caucasian one is less known. Edward crossed and recrossed the Caucasus, but only by the main road from Tiflis to Vladi Kaukaz.

It is very difficult to penetrate to right or left of that one pass, the chain of mountains being

1872. singularly continuous and unbroken. The houses of
the peasantry in these mountains appeared to him
remarkable, as having nothing of the usual character of mountain dwellings. They are flat-roofed
and not different in appearance from those in the
lower country. The people are in the habit of
setting up their haystacks on these flat roofs, which
has a very odd look. Edward tells me that some
years ago, when he was travelling in Hungary and
Transylvania, he saw eight eagles at once, golden
eagles he believes, which rose from the carcase
of a horse on which they had been feeding; while a
huge vulture, which had been engaged in the same
way, lingered some time longer before he took
wing. This was in the mountains between Transylvania and Wallachia.

Tuesday, January 16th.

We walked round the grounds with our guests.
I went with the Bishop to a Church Defence
meeting at Bury—very tiresome.

The Barnardistons and Lady Barbara Legge,
and Cecil arrived.

Lady Cullum, the Hardcastles, Mr. and Miss
Bevan, dined with us.

Wednesday, January 17th.

Excessively wet and stormy.

The Bishop and Mrs. and Miss Browne went
away after luncheon.

Miss Evelyn Bevan and her brother dined with
us.

Thursday, January 18th.

A sad disappointment about Mrs. Walrond. I wrote to her.

Fanny went with the Barnardistons to Elvedon, to call on the Maharanee.

All our guests went away except Cecil.

————

Friday, January 19th.

We dined (including Cecil) with the Wilsons— met the Milbanks, Lady Cullum, the Robert Anstruthers, Mr. Powell, John Hervey and others.

————

Saturday, January 20th.

Cecil went away—Frank Lyell arrived.

————

Tuesday, January 23rd.

Mrs. Wilson and her daughter and niece dined with us, and went with us to the Bury Athenæum— Lord Arthur Hervey's excellent lectnre—delighted to meet with him and Lady Arthur and Sarah —we met also Lord Charles and Lady Harriet, Lady Mary, &c.

The lecture was on the different versions, ancient and modern, of the Hebrew Scriptures, from the Septuagint down to our own times. It was very good and less dry than I had expected from the subject; and it was very interesting to sit again listening to a lecture from *him* in the old place, just as in the old days before he was removed to Wells.

1872.   After the lecture, we had the great delight of greeting our dear friends—the Bishop himself, and Lady Arthur and Sarah. They had arrived the same day at Ickworth.   Kate is with her brother at Shotley.

Wednesday January 24th.

Frank Lyell went away.   He has been spending three days with us, a farewell visit, preparatory to his setting out for La Plata ; for he is to start on the 10th of February.   I heartily wish him success, he is an interesting and pleasing young man.

A furious storm of wind this morning, which shattered some old trees, but did less mischief than I expected, and in the arboretum (I am glad to find) no damage at all.

Thursday, January 25th.

A public luncheon at the Angel on the occasion of presenting to Lady Arthur Hervey a portrait of her husband (by *Graves*) for which a number of persons in the district had subscribed.   I was in the chair ; Lady Arthur on my left, Lord Arthur on my right ; Sarah, Kate, Patience and John Hervey, Fanny, Lady Cullum, Lord Charles Hervey, Archdeacon Chapman, Mrs. Wilson near us.

My speech was successful, and indeed I think was better than most of my performances in that way. Lady Arthur spoke a few sentences of thanks with great simplicity and feeling.   The Bishop's speech admirable, as is usual with him.

After the speechifying, he and I reminded each other of the beginning of our acquaintance, when in

the autumn of 1829, and on the eve of my first 1872.
going up to Cambridge, we met and were introduced
to each other at the house of Mr. Hasted.

Sunday, January 28th.

We have been grieved by the death of Miss Ella
Kennaway, Sir John's only daughter, a very interesting and attractive girl, handsome, clever, accomplished and agreeable. We have heard no
particulars, and only know the fact from the
newspapers.

Monday, January 29th.

Our dear friends the Arthur Herveys, the Bishop,
Lady Arthur, Sarah, Kate and John came to stay
with us; also Mr. Charles Hoare, Kate's betrothed.
Patience had already been staying with us since the
25th.

Tuesday, January 30th.

Very mild. A delightful walk with the Herveys
about the grounds: in afternoon they drove to
Stowlangtoft and Langham with Fanny. Another
very pleasant quiet evening.

Wednesday, January 31st.

A beautiful day, perfect spring. Another very
pleasant walk with the Herveys; we planted a tree
for Kate and Mr. Hoare.

The Chapmans, Mr. Abraham, Mr. Lott, Mr.
James and Mr. Shaw dined with us. Fanny has
since had an interesting and very touching letter

1872. from poor Sir John Kennaway himself, giving a particular account of his daughter's death. She had been ill for some days with a cold and rheumatism, but no danger was apprehended, and her death was quite sudden :—caused, as it is supposed, by the rheumatism attacking her heart.

The weather extraordinarily mild through almost the whole of this month. No snow at all, and no frost except in one or two nights, scarcely at all, I think, during the day. Frequent and heavy rains (very good for agriculture), and great storms of wind. This day (the last of January), has been really beautiful—mild and fine enough for April.

* * *

Thursday, February 1st.

We rambled about with the Herveys and Anstruthers, and choose a place for planting a tree.

Edward arrived.

Mr. and Mrs. Thornhill, Mr. and Miss Rodwell, Mr. and Miss Bevan dined with us.

* * *

Friday, February 2nd.

We planted a tree (a Cephalonian fir*), for the Bishop and Lady Arthur : afterwards we had a delightful walk with the whole party.

Lady Cullum, Patrick Blake, the Wilsons, Miss Bevan and her brother and Mr. Wratislaw† dined with us.

* * *

* This tree is planted in the Park, close to the path to the Vicarage grove.

† Master of the Bury Grammar School.

We planted two trees in the Dairy Grove for Sarah and Patience Hervey.

The Barnardistons and Lady Barbara, the Anstruthers, John Hervey and Alfred Newton went away. Sir Frederick and Lady Grey, G. Napier, young Henry and Clement arrived.

———

Sunday, February 4th.

My 63rd birthday—thanks be to God.

We went to morning Church and received the Sacrament with the Arthur Herveys.

———

Monday, February 5th.

Lord Arthur Hervey went away early. Lady Arthur and the girls after breakfast. Edward and Clement also went away. We took a walk with the Greys. The two John Herveys and Mr. Bowyer arrived. We have had a truly delightful week of the company of our dear Arthur Herveys : nothing could be more enjoyable. Very sorry indeed I was to part with them this morning.

═══

LETTER.

Barton,
February 5th, 1872.

My Dear Mary,

Very many thanks for your kind note on my birthday. I am exceedingly glad to hear that you are feeling so well and so brisk and young; long

1872. may you continue to feel so.   For my part I have
nothing to complain of, but much to be thankful for.
I never feel very *robust* in winter, but I have nothing
particular the matter with me, indeed I am in very
tolerable health and as happy as can be.   I trust
that Lyell also is in good health.

I have no doubt your thoughts are at present much
occupied by Frank's approaching departure ; I hope
and trust all will turn out well for him, and that he
will like La Plata, and profit in many ways by his
visit to it.   I always think it is a good thing for a
man to see a variety of countries and *peoples* (to
adopt the modern plural) while he is young.

I was very much pleased with Frank when he was
here a little while ago ; he is a pleasing and
interesting young man, very intelligent and seems to
think for himself, and I think there is every probability of his being a good and wise man.

I have lately looked through John Phillip's
" Geology of Oxford," which seems very well done;
only it appears to me rather eccentric to include the
Malvern Hills in the valley of the Thames.   Also
his *Poikilitic Cainozoic* nomenclature apppears to me
affected and disagreeable : on the same principle he
ought to say Eocaine and Miocaine.   It might be
very well for Mr. Grote, in the actual old Greek
names, to imitate the old Greek spelling (though
even he is not quite consistent), but in modern
scientific nomenclature I like to stick to the
established practice, of using Greek *filtered* through
Latin.

We have just parted with our dear friends the

Arthur Herveys, having had a delightful week of 1872.
their company; and it is very pleasant to see how
happy they all are in the prospect of Kate's
approaching marriage. Sir Frederick and Lady
Grey are here, also Edward, but he goes back to
town to-day: also my nephew, young Henry (or
"Henny"), who is still exceedingly handsome and
very pleasing and gentlemanlike.

Fanny is well, and I need not say, busy.

With my best love to Lyell,

Believe me ever,

Your affectionate brother,

CHARLES J. F. BUNBURY.

JOURNAL.

Tuesday, February 6th.

Minnie Powys arrived. Lady Cullum, Patrick
Blake, Captain and Mrs. Ives, the T. Thornhills
and two Miss Bevans dined with us.

Wednesday, February 7th.

We walked about the grounds for some time with
the Greys, Minnie Powys and the two Herveys.

I read part of young Henry's journal.

General and Mrs. and Miss Eyre and Mr.
Milbank arrived.

Thursday, February 8th.

We planted a tree for Sir Frederick and Lady

1872. Grey. A visit from Mr. James. Leopold Powys arrived.

The Greys, Leopold Powys, Mr. Bowyer and John Hervey went away. We walked with the Eyres and Minnie Powys.

The Eyres went away in the morning, Minnie Powys and George Napier in the afternoon.—Henry only remaining with us.

Sir Frederick Grey thinks that among the misfortunes of France, one of the greatest is the having Thiers at the head of affairs. He quite agrees with me, that Thiers' Histoire du Consulat et de l'Empire had a great share in producing and fostering that state of feeling among the French which led them into the last war: the notion that it is their right and their province to dictate to, and to domineer over the rest of Europe.

Sir Frederick Grey thinks that the Americans will not press their demands as to the Alabama claims, but have merely put them forward as an experiment to try how much we may be willing to concede. He, however, and all our friends, are much impressed by the serious importance of the crisis. It is, indeed, *the* political question of the time.

General Eyre has much knowledge, observation and thought, and his conversation is well worth attending to. George Napier is always very pleasant and full of good nature, good humour and good sense.

————

Tuesday, February 13th.

Consultation with Scott and Allan about the fern-house and about a new path.

The dreadful news of the assassination of Lord Mayo, the Governor General of India.

====

LETTER.

February 14th, 1872.

My Dear Susan and Joanna,

I thank you both for your very kind notes and good wishes on my birthday, and assure you it gave me great pleasure to receive such tokens of your kind remembrance. We did not go to church *tête à tête* on my birthday, but with a large party— five Herveys, Sir Frederick and Lady Grey, Edward, Clement, young Henry, little Arthur and his governess.

I have begun to read through Grote's " History of Greece " (passing over the mythology, however), a good long job ; as yet 1 have got only to the time of Solon, and so far I find it intensely dry.

Now adieu, dear sisters.

Believe me ever your affectionate brother,

CHARLES J. F. BUNBURY.

====

JOURNAL.

1872.   Fanny with Henry went over to Mildenhall, and returned to dinner.   I went on with my herbarium catalogue and arrangements.

Lady Grey, while here, told me a good deal about that wild and lonely valley in Norway, where she and Sir Frederick Grey spent some time in the summer of '69, much to the same effect as what I wrote down from her letter, in my journal of July 28th, '69. She said that it was like living at the bottom of an enormous trench : the mountains rising almost precipitously to a great height on each side; the sun visible only from about 9.30 in the morning to about 3.30 in afternoon ; but there was no night —no darkness—daylight or a clear twilight lasted all the twenty-four hours.

Two of the most common plants in the valley were Cornus Suecica and Trientalis Europæa; Lilies of the Valley also in abundance, and Linnæa borealis.

Walked with Fanny to look at the new path, &c.

Walked with Fanny and Henry.

Lady Cullum came to luncheon, and we showed her the arboretum, farm, &c., and walked about with her for nearly three hours.

Saturday, February 24th.

Up to London with young Henry, to St. Pancras station—a prosperous journey.

Minnie and Sarah and Henry came to see us in the evening.

February 25th.

Great preparations making in London for the ceremony of the 27th,* and much excitement.

William Napier showed us some of the apparatus of the *Kriegs-Spiel* (the new German device for the study of tactics) which he and General Eyre have introduced into England ; and explained to us the method in which it is worked—for it appears to be much too serious to be called play.

Februrary 26th.

A pleasant visit in the morning from Charles and Mary Lyell ; Mary looking *very* well ; Charles, I am sorry to say, not in a comfortable state of health, though certainly better than he was last summer.

Charles Lyell strongly recommended to me the last article in the present number of the *Quarterly*, " The Proletariat on a False Scent." I have since read it, and I think it admirable ; only I dislike the word *Proletariat*—a piece of French political slang.

* The Thanksgiving at St. Paul's for the recovery of the Prince of Wales.

1872. Afterwards, while at luncheon, we had the great and
unexpected pleasure of a visit from Charles Kingsley,
whom I had not seen for (almost exactly) three
years. His daughter Mary was with him. He has
come to town, of course, to attend the Thanksgiving
service. He stayed but a short time, but talked
very agreeably, quite in his old way, and seemed in
great force, as frank and cordial, as if it had been
only a week since our last parting; and I thought
him looking very well. He spoke with high satisfaction of the progress which his son Maurice is
making in the far West of America — of the
reputation he has established, and on the money
he is making; talked also much of the wonderful
natural features of that country, particularly the
canons or gorges and the enormous water-worn
rocks. He talked also of Trinidad and his interesting visit to it. He said what I can well understand,
—that he felt overwhelmed with the multitude of
new impressions; that though he began his notes
from his first arrival and kept them every day as
diligently as he could, yet he found that the
multitude of new things to be noted continually
outran his power of recording them, and day after
day he felt the insufficiency and incompleteness
of his memoranda. And not only so, but he felt
not less the insufficiency of his powers of attention
and observation. Continually, he said, it happened
that before the day was over, he felt that his
power of attention was worn out, and he could
observe no more. I quite understand and sympathise with him.

Afterwards we called at Mr. Hoare's in St. James's Square, and had the delight of seeing dear Sarah and Kate Hervey, who are staying there till the 28th—both looking very well.

We wound up this agreeable day by drinking tea with the Charles Lyells, with whom we again had a pleasant talk.

Lyell gave me the first volume of the new (11th) edition of his "Principles," which has just come out. His devotion to those subjects is as strong —perhaps even more intense than ever; and his intellect as clear as possible; but constant care and watchfulness is needed—and bestowed by his wife that he may not overwork and fatigue his brain.

Tuesday, February 27th.

A fine day, but cold wind. The Thanksgiving ceremony—Fanny saw the procession well, from Wilkinson's in Pall Mall. I did not go. In afternoon we visited the William Napiers, and also the Walronds, who were very pleasant.

Thursday, February 29th.

Leonard Lyell came to luncheon with us. General Eyre also called. Edward and Henry dined with us.

The great ceremonial of the 27th, the Thanksgiving, was a most complete and satisfactory success; very gratifying to the Queen and her family, and to all good subjects, and very honourable to the people.

1872. All the people I have seen who witnessed it, agree that nothing could surpass the good humour and good behaviour of the vast multitude, and that the welcome everywhere given to the Royal Family was evidently heartfelt. Such a scene must do good. I had a threatening of cold and face-ache, so stayed at home. Yesterday, the 28th, we visited the Exhibition of Pictures by the Old Masters, at the Royal Academy Rooms, Burlington House; a beautiful collection. I had sent two of our pictures (Mrs. Bunbury and Mrs. Gwyn, by Sir Joshua Reynolds), which I now saw very advantageously placed. Some of those I particularly admired were —"Nelly O'Brien," *(Reynolds); Rubens's* "Rainbow Landscape," (these two from Sir Richard Wallace's gallery), *Rembrandt's* "Portrait of his Mother," and of "A Burgomaster's Wife;" a magnificent portrait of "An unnamed Man," by *Rubens;* a landscape ("The Timber Waggon,") by *Rubens;* "A Portrait of a Lady in a blue dress" (without a name), by *Reynolds;* and "Portrait of Francis Hare," by the same; and *Danby's* "Calypso." "Portraits of Mrs. Hoare and Child," by *Reynolds.* "View on the Thames at Mortlake," by *Turner.* "The Cherry Seller," by *Collins.* "Portrait of Miss Linley," by *Romney.* And a wonderful "Portrait of a very old Lady reading," by *Rembrandt.* "The Portrait of Clarendon" is interesting for its subject. *Etty's* "Triumph of Cleopatra," is a very fine specimen of the artist. *Rembrandt's* "Portrait of the Wife of the Burgomaster Palekan" is quite wonderful for the finish and the expression of the face.

From thence we went to Harley Street and had 1872. luncheon and a pleasant chat with the Charles Lyells. Lyell was in good spirits, and I thought him looking better than on the 26th.

I heard however afterwards that he was fatigued by the excitement of conversation on this occasion.

Afterwards, we drove a little way along Oxford Street, to see the gay and fantastic decorations of the shop fronts, which had been put up for the procession. London hardly looks like itself, rather like an Italian city on some grand festa. Leonard Lyell came to luncheon; he is a young man of great acquirements and great ability.

In the afternoon we called on and saw our dear friend Lady Bell, now in her 85th year and still in fine preservation, with considerable remains of beauty, with her faculties perfect, her sweet, winning manners and equally sweet nature unaffected by age.

Friday, March 1st.

Talk with Godwin Austen at the Athenæum.

Visits from Mary, Henry and William Napier. We dined with the Walronds—met nobody but Miss Moore and Sir Edward Ryan: but it was a most agreeable evening, thanks to the host and hostess. Mrs. Walrond is as fascinating as ever, and Mr. Walrond exceedingly pleasant as well as most estimable.

1872.   Saturday, March 2nd.

Drove out in the morning with Fanny and young
Henry. Octavia and Wilhelmina Legge came to
luncheon. William Napier and Leonard Lyell
dined with us. We visited the National Gallery,
enjoyed the Peel pictures and saw the Raphael (a
Holy Family with Angels), which is not, indeed,
definitively added to the gallery, but it is for the
present exhibited here, being offered to our Government for (it is said) £40,000. A large price,
certainly, and more than I can imagine the picture
is fairly worth. It is in Raphael's very early style,
almost exactly like that of Pietro Perugino. A
peculiarity in it is that the Holy Infant is completely
draped, which is said to have been done in com-
pliance with the ultra-delicate scruples of the nuns
for whom the picture was painted. We both
thought the famous "Chapeau de Paille" rather
inferior to Reynolds's "Nelly O'Brien," which we
had lately seen at Burlington House.

                                       Sunday, March 3rd.

We sat some time with the William Napiers,
then called on Mrs. Young and walked home across
the Park.

George Napier, Edward Clement, Herbert dined
with us.

                                      Monday, March 4th.

A very beautiful and quite warm day, perfect
spring. Went with young Henry to the Jermyn

Street Museum, and looked through part of the 1872.
collection of minerals with him.

We dined with the William Napiers, George also
there.

———

We returned home by the 4.25 train from Shore-
ditch, by Sudbury; reached Barton before 8. All
well—thank God.

My nephew, Henry, left us in the morning of the
5th, to return to Abergwynant, having been with us,
at Barton and in London, from the 3rd of February.
I am extremely pleased with him. He is singularly
gentlemanlike and pleasing in appearance and
manners, courteous and gentle to a degree not usual
in lads of his age, very amiable and affectionate,
good principles he could not fail to acquire from his
father and mother. Intellectually, perhaps his
education has not been quite so well attended to in
some respects, as might have been wished : his
writing is very imperfect, and he is almost ignorant
both of Latin and French. But he has a most
decided and remarkable turn for scientific studies,—
indeed, I may say a devotion to them, which
pleases me very much ; he seems to be never tired
of studying books of that kind, and I am in hopes
that he has a real talent, a real vocation for pursuits
of that kind, which will be of the greatest advantage
to him in many points of view. He is now ordered
out again to the Mediterranean.

———

Wednesday, March 6th.

Barton.

Very bright and clear with very sharp E. wind.
A long talk with Scott. Looked round the gardens,
&c.

Thursday, March 7th.

Went to look at the newly planted trees.

Friday, March 8th.

Strolled with Fanny and two dogs.

The sudden, though I cannot say premature,
termination of the Tichborne case is striking. The
villainous Claimant is now in prison, to take his
trial for perjury; I hope he will be convicted and
punished with the utmost rigour of the law.

March 11th.

Our two pictures—*Reynolds's,*—Mrs. Bunbury and
Mrs. Gwyn, which we have lent to the Royal
Academy for exhibition, have arrived, were un-
packed on the 11th, and found to be safe and
unhurt.

Tuesday, March 12th.

My Barton Rent Audit — very satisfactory:
luncheon after it with the tenants and Mr. Smith.

Wednesday, March 13th.

We had a large Thuia removed from the
arboretum into the grove.

Mr. Allington, the School Inspector, arrived. Lady Cullum and the Percy Smiths dined with us.

Friday, March 15th.

Little Arthur's birthday. Mr. Allington went away after inspecting our schools.

Monday, March 18th.

Lady Bristol brought Mr. Graves, the artist, to see our pictures ; both were very pleasant.

Tuesday, March 19th.

The Quarter Sessions at Bury—large muster of magistrates.

Wednesday, March 20th.

It is certainly honourable to the Italians, that they have shown such general respect for the memory of Mazzini* though he was so violent a partizan, and so bitter an opponent of the party (the Piedmontese and Monarchial) which is now triumphant.

March 25th.

A very heavy fall of snow yesterday morning, lasting till past 10 a.m. I have not seen so much snow at Barton for very many years.

Resumed the work of revising my father's Hawkwood MSS.

* This extraordinary man died at Pisa, on the 10th of this month.

Thursday, March 28th.

Began to read Sir John Hanmer's book of Family
History.

_____

Monday, April 1st.

We dined with Lady Cullum—met Lord Talbot
de Malahide, Mrs. Ives, Lord John Hervey,
Beckford Bevan and his daughter, and the Percy
Smiths.

_____

Tuesday, April 2nd.

Fanny went with Arthur and Miss Hamilton*
to Mildenhall and returned to dinner. Scott gave
me a satisfactory report of the Mildenhall School
Board.

Much grieved by the news of the death of Mr.
Maurice ; an admirably good, and most interesting man of a very high order of mind. I
did not know him so well as I could have
wished, though it is many years since I first
made his acquaintance ; and he has once or
twice been our guest here ; but his manifold
occupations, his wife's wretched health, and perhaps
in some degree his own, prevented him from accept-
ing many invitations,   I had a very great respect
for both his intellectual and his moral character.
His manners were very pleasing, peculiarly mild and
unassuming. He is a very great loss, both publicly
and privately.

I well remember Kingsley saying to me, many
years ago, that he thought Mr. Maurice the very
best man he had ever known.

* Miss Hamilton was Arthur's governess.

Up to London. We dined with the William Napiers: a family party of themselves, Sarah Craig, Edward and Cecil. A pleasant evening.

———

We dined with the Charles Lyells.

— — —

Torquay.

We went down to Wells by the express train on the 8th; Lady Harriet Hervey in the same compartment with us. On the 9th dear Kate Hervey was married to Mr. Charles Hoare. The ceremony, in the glorious choir of the Cathedral, was a beautiful and most interesting sight. All the arrangements were perfect; Kate in her sweet innocent look of blushing happiness was lovely, and the effect of the scene when she and her husband and her bridesmaids were kneeling was quite touching. The ceremony was performed by the Dean, Dr. Johnson. and the arrangements in the Cathedral, I believe, were made by him and his wife. The bridesmaids were a pretty group. There were ten of them:—Kate's three sisters, Lady Mary Hervey, Miss Rodney, Miss Isabel Hervey and four Miss Hoares, sisters of the bridegroom.

The wedding breakfast, at which nearly two hundred persons were present in the noble crypt or "undercroft" of the Palace, was also an interesting

1872. scene; and some of the speeches were good; the Dean's in particular, capital. (His health was proposed by the Bishop, as he had performed the ceremony). The Bishop spoke, as he always does, with fine taste and true feeling. Altogether it was a most interesting and memorable day—the beginning I hope and trust of a long career of happiness for dear Kate and her husband. I am very sure no element of goodness or wisdom will be wanting on her part.

We visited Lady Sidney Osborne, after the wedding on the 9th, and saw her and Lord Sidney; the next day she and her daughter called on us at the Palace, and talked with us some time. I took a great fancy to this Miss Osborne.

We remained at Wells during the 10th and 11th. We had great pleasure in renewing and somewhat improving our acquaintance with Lady Sidney Osborne and her daughter; and I was very glad also to become acquainted with the celebrated Lord Sidney Godolphin Osborne (S.G.O.).

He is a very interesting man; striking and impressive even in his appearance, and more so in his conversation—or rather indeed, discourse,—for it was hardly *conversation*. He launched out into his favourite topic on which he has thought and written so much—the condition of the peasantry; and he talked on it I should think fully half an hour, with great fluency and command of language, with much earnestness (but not vehemence) and great power of expression. I was much interested. Lord Sidney Osborne is of course deeply interested

in everything connected with the labourers' strike.
He feels a warm sympathy with them, and he has
some hope that this new spirit of combination may
have a good result, *if* the upper classes do not
attempt to check it by harsh measures, and if the
regular political agitators and stump-orators of the
towns do not manage to convert it into political
capital. He said he should at once place himself in
decided opposition to it, if he saw it likely to
degenerate into a political agitation. He said that
it would be a great, though silent revolution in the
country, if the peasantry generally came to lose
their strong local attachments: that hitherto, the
English labourer's *patriotism* had been *parochialism*,
that all his strongest feelings and instincts had been
bound up with his parish, and with the employers
under whom he and his fathers had lived : and that
if this were altered, if he came to consider it a
simple and natural thing to migrate in search of
better wages, the result would be a general *unsettling*
of which no man could foresee the end. Lord
Sidney said, that he was satisfied that it had been
usual to overrate very much the profits derived from
agriculture—both the farmers' and the labourers'
share.

Among the multitudinous wedding party, we
made an acquaintance with Mr. and Mrs. Locke
King, and found them very pleasant. I had not
met Locke King since our Cambridge undergraduate
days.

We left Wells on the 12th, and came to Torquay ;
the country is beautifully varied, exquisitely green,

1872. and enriched like a garden with the blossoms of innumerable fruit trees ; the coppices and hedge-row trees in all the beauty of their young foliage, and the green banks everywhere enamelled with primroses. A lovely country and a lovely season. In this tract of country, especially as we approach Exeter, the bright red (sometimes almost *vermilion*) colour of the soil, the brilliant green pastures and the dark red cattle are striking features. These Devonshire cattle, dark red, with long horns, seem as characteristic of the country as when I first was here in '26. Torquay looked very bright and attractive, with the blue sea and white houses sparkling in the sunshine, and we were soon comfortably settled in a first-rate hotel.

We remained at Torquay from the 12th to the 16th, and left it with much regret. The weather all these days was perfectly delicious, warm and bright and sunny enough for the middle of June. The picturesque ground, the rocky hills, the gay white houses, the rich foliage and the beautiful sparkling bright blue sea, reminded us strongly of the Mediterranean coasts : but there is a much greater prevalence of the brilliant *greens* of spring— of fresh grass and young leaves, than one usually sees in the South.

On the 13th we drove out with those two very interesting girls, Mary and Theresa Boileau, along the coast to the northward ; Theresa, I grieve to

say, is in very bad health; we drove to Ansty's 1872.
Cove, Babbicombe, Mary Church and Watcombe—
a very agreeable drive. The day superb, and the
coast very beautiful. The marble rocks and the
general style of the coast of Ansty's Cove in
particular, reminded us forcibly of Monaco, Villa-
franca and other spots near Nice.

April 15th we drove to Kingswear, opposite to
Dartmouth, in order to see that harbour, which was
new to me. The day was beautiful and the drive
very agreeable. Dartmouth is certainly a very
remarkable spot—the harbour is formed by an
expansion of the river Dart, a little way above its
mouth; the hills closing in round this basin on
every side, and hemming it in in such a way that
one can see no outlet; and the harbour has perfectly
the appearance of a small lake, completely inclosed
by steep hills. A port more thoroughly land-locked
cannot well be.*

Leaving Kingswear, and ascending some way up
the hill behind it, and going a few yards off the
main road, we looked down on the actual mouth of
the river, an interesting spot; the Dart contracted
to a much less width than where it forms the
harbour, and winding between the hills which
appeared to bar its passage, escapes into the sea
between two abrupt rocky points.†

The 16th we left Torquay, journeyed to Bath.

---

* " Portus ab accessu ventorum immotus,' most completely ;—but scarcely
" ingens."

† It appears by the map that the mouth is not really much narrower than the
harbour, though it looks so.

1872. From Bath we drove in the afternoon to see Mr. and Mrs. Drummond, at St. Catherine's, beyond Batheaston : a very pretty, picturesque, interesting old house, originally (I believe) a monastic building, which they have fitted up with great taste.

April 17th from Bath to London : arrived safe and well (thank God) at Eaton Place.   Henry, Cissy and Emily came to see us.

Went to Ebury Street and voted for Lord Mahon for the School Board, then drove with Fanny to Lambert's and bought a diamond ornament, then to see the French pictures.   Henry, Cissy, Emmy and Edward dined with us.

My dear brother Henry, yesterday morning, underwent a most serious operation, performed by Sir James Paget.   I saw him (Sir James) late yesterday afternoon, and he assured me that Henry was going on as well as possible.

What a blessing the invention of chloroform has been to mankind!

Henry going on well—I sat with him some time. Katharine, Harry, Leonard and Clement dined with us.

Lady Grey came to luncheon with me (Fanny having gone to the Drawing Room), she was very agreeable. She said she hears that the reply or " Counter-case " drawn up on behalf of our Government in the American controversy is quite a masterpiece—perfectly conclusive. It seems to dispose effectually, not only of the *indirect claims*, but of the direct claims also. Its principal author is understood to be Sir Roundell Palmer.

Lady Head and her daughter Amabel, William and Emily, George, Edward and Emmy dined with us—a pleasant party.

Sarah and Patience Hervey came to luncheon with us : dear Sarah looking well, and charming as ever, seemingly quite recovered from the fatigues of the wedding festivities at Wells.

We went to Veitch's nursery garden in the King's Road—a wonderful collection of rare and beautiful exotics. A whole house full of Nepenthes, of many species and in all stages of growth, most curious and interesting. Dionæa in perfection under a bell glass, Sarracenia, of various species, in fine condition. A vast collection of exotic ferns, in the highest beauty of condition, especially several species of Trichomanes and Hymenophyllum, which are so difficult to cultivate in general : Trichomanes reniforme very fine, Trichomanes speciosum in as rich luxuriance as we saw it in Teneriffe, and many

1872. others. Leptopteris superba and pellucida, which require the same treatment as the Hymenophylleæ, are here most beautiful. Many noble tree ferns.— An exceedingly curious little fern, which I never before saw alive, was Rhipidopteris peltata—formerly called Acrostichum.

———

April 25th.

Delightful news ;—our dear Sarah Napier is engaged to be married to Lord Albert Seymour, the second son of the present Lord Hertford, and the match appears to make her dear mother very happy. This news was indeed made known to us last Monday, the 22nd, in a letter from Minnie to Fanny, but it was then a secret. This day Minnie and Sarah came to town, and the preliminaries of the treaty having been settled, Minnie immediately came to see us and to confirm her first announcement. Both mother and daughter seem thoroughly happy, and this makes me too very happy, for I have long loved Minnie quite as my sister, and Sarah almost as my daughter, and any good which happens to them is delightful to me. Sarah is not only lovely but loveable in the highest degree, and has the sweetest temper and disposition that can possibly be : I am sure she has every quality that can conduce to happiness in marriage. We hear very good accounts of the young man, and of his father and mother : he seems to be of a very good family in every sense.

———

Weather very fine.

We visited Minnie and Sarah—afterwards went to a very pleasant afternoon party at Mrs. Hoare's.

April 27th.

We are very anxious about our charming friend Mrs. Walrond, who has been dangerously ill, and I fear is not yet out of danger, though the last accounts are decidedly better.

Henry is going on well, but his recovery is slow.

April 28th.

We had yesterday a little dinner specially for the *betrothed*,—to improve our acquaintance with Sarah's future husband.

We formed a very favourable opinion of Lord Albert. All that we hear, from a variety of quarters, about his father and mother, is most agreeable, and encourages the best hopes for dear Sarah's happiness.

Mr. Clements Markham, whom I met at Lady Mary Egerton's, told me that the Chinchona trees on the Andes always grow on very steep slopes, on the steep sides of mountains or in ravines where torrents flow rapidly; what they cannot bear is stagnant water about the roots. In the Andes they are very local, and the special districts of different species are often separated by hundreds of miles ; but in the plantations in India, where several distinct species are cultivated together, they show a great disposition to hybridise. He remarked that

1872. the genus Chinchona, as now limited by botanists, is strictly confined to the Andes ;—that the character which had been thought absolutely distinctive of the true medicinal Chinchonas (the capsule opening from the base upwards), is now found not to be without exceptions.

Mr. Clements Markham spoke of the vast materials for illustrating the botany of Spanish America,— heaps of dried plants, elaborate drawings, and notes, —collected by the various scientific expeditions sent out in the reign of Charles III., and still lying idle at Madrid ; for only a small part of them were worked up into the great but incomplete work of Ruiz and Pavon. The collections of Mutis (Linnæus's correspondent) remaining unexamined at Madrid, were immense. Nothing has been done by the Spanish Government, since the beginning of this century to *utilise* these materials—and they are not willing to let foreigners do it for them. He said he believes that the climate and vegetation of Spain, and indeed of parts of Southern Italy also, have been much altered within the human period, by the reckless destruction of the forests. Owing to this destruction, not only the ground has been burnt up by the direct rays of the sun, but much of the vegetable soil has been washed away by the storms ; and many tracts formerly covered with forests, will now grow nothing but such plants as the cactus and agave, which require but little soil. The Spanish Government is now endeavouring to plant and re-establish forests, but Mr. Markham thinks that much difficulty will be experienced.

A very pleasant visit from Sir Frederick Grey.

Happily Mrs. Walrond seems to be making satisfactory progress towards recovery.

Went with Emily Napier and Fanny to see the
Water Colour Exhibition (the "Institute," near the
west end of Pall Mall). We were extremely pleased
with it: it is one of the best exhibitions of the kind
that I have seen. Some of the pictures that I
particularly noticed and admired were: "The great
Bazaar at Cairo," by *Carl Werner;* "A View of
Benares," by *T. L. Rowbotham;* "The Ford" (a
body of horsemen in the costume of the 16th
century, fording a stream), by *Cattermole;* "Jonas
Hanway and his Umbrella," (very humorous and
natural), by *J. D. Linton;* "The Hall of Justice of
the Franc de Bruges," by *Louis Haghe;* "Harting
Combe," by *G. Shalders;* "Wargrave on the
Thames," by *J. H. Whymper;* and some beautiful
Flower-pieces, by *Mrs. Duffields.*

May 4th.

On the 1st, Henry was sufficiently recovered to be
able to dine with us. Later in the evening, I went
to Sir Henry Rawlinson's "Reception," (as President
of the Geographical), in Willis's Rooms:—Cissy
and Emmy went also with Edward. It was a large
assembly of ladies and gentlemen. In the middle
of it Sir Henry Rawlinson stood up on a chair and
read aloud the telegraphic announcement, which
had just been received, of Livingstone's safety; this

1872. news was loudly cheered. Those whom I knew in this party were :—Lady Mary Egerton and two of her girls; Sir Bartle and Miss Frere; Lady Colebrooke; Admiral Spencer; Mr. Twisleton ; Mr. Murray ; Colonel Pinney.

On the 2nd, we came down to Barton in the afternoon, arrived at our dear home about 7.30, and found Arthur and all well. Thank God. Those first two days of May were beautiful and quite warm ; really worthy of the poet's May. Yesterday, the 3rd, also very fine, but in the afternoon the wind turned cold. This place is in great beauty, and the season an early one. Lilacs in the richest profusion of blossom ; horse-chestnut also in full flower ; laburnum beginning ; the Judas tree in bud ; magnolia conspicua and purpurea in great beauty ; also pæonies, the great iris, and a variety of herbaceous plants. The grass very rich ; the oaks coming into leaf, and the beeches in all the beauty of their young foliage. We have heard the nightingale, but not yet the cuckoo.

————

May 6th.

Walked to see the newly-planted trees.

————

Wednesday, May 8th.

We went with Arthur to the gravel pits and strolled about.

There were some heavy thunderstorms here yesterday : one in particular, in the afternoon, very violent ; and I hear that an old oak tree on Lofts's

farm, not far from the Boys' School, was shattered 1872.
by lightning. At the same time the weather was
very cold.

— ·—·—

Ascension Day. We went to morning Church
—afterwards strolled about the arboretum, &c.
with Arthur.

Mr. and Mrs. Montgomerie came to luncheon,
and spent the afternoon with us. Young Arthur
Hervey and Francis Anstruther, also called.

Strange weather : storms of cold rain, hail, sleet,
and once (between seven and eight in the morning),
actual *snow*, accompanied with excessive cold ; yet
between these storms there were bright and even
warm gleams of sunshine.

Very fine. Walked with Fanny and Arthur—
looked at the newly-planted shrubs. The day
before yesterday we received news of the death of
our dear and charming young friend Mrs. Walrond.
I have not for many years heard anything which has
made me so unhappy. I have rarely known a more
fascinating person. Whether in her gay or in her
earnest moods, she was equally delightful. And we
have every reason to believe that she was as good as

1872. she was charming. And all these rare gifts, all this beauty, this fascination, these talents, this power of making others happy, are buried in an early grave, (she was only 31), while so many who are neither useful nor ornamental live on cumbering the earth to extreme old age. Truly it is mysterious. Much as we mourn for her, our sorrow must be as nothing compared to that of her poor husband, who adored her; I dread to think of his desolation. He is left with six children, the eldest hardly 12 years old :— lovely children they are, those that I have seen of them, and they will relieve his sense of loneliness, though they increase his cares.

Friday, May 17th.

We spent all the morning in examining MSS. and comparing them with Mr. Horwood's Reports.

Saturday, May 18th.

Wrote to the Secretary of Historical Manuscripts Commission. Fanny had a charming letter from Kate Hoare, from Venice. Katharine Lyell writes to me that she has had a letter from Lady Smith, the widow of Sir James Edward Smith, who is now in her *hundredth* year. Katharine wrote to congratulate her on her 99th birthday, and says that Lady Smith's letter in reply was a very good one.

LETTER.

My Dear Katharine,

I thank you very much for your letter, and most especially for the full information about dear Frank. I was very glad to hear of Charles and Mary's safe arrival, and that they are looking well, and seemed to have so much enjoyed their tour. I am much interested by what you tell me of old Lady Smith; what a wonderful woman she is? I hope she will be spared to complete her century of life.

I am grieving for the death of Mrs. Walrond; it is long since anything has given me such a shock. I have seldom known a more charming person. We had taken to her exceedingly from the very first, and our regard for her went on increasing rapidly as we knew more of her; and I had so looked forward to frequent future visits here from her, and to cultivating her friendship as long as I lived. I am sure Lady Bell will quite sympathise with me. I have had a most touching letter from poor Mr. Walrond, to whom I had written. *She* was to be buried to-day.

I quite agree with you that Mr. Kingsley ought to rest his brain, and I *hope* he will. I do not feel happy either about him or Mrs. Kingsley.

With much love to Harry and your children,

Believe me ever,

Your affectionate brother,

CHARLES J. F. BUNBURY.

1872.

JOURNAL.

1872.   A fine day.

The Barton labourers' club feast, I made a short address to them.

———

Mr. and Mrs. Montgomerie arrived.   Mr. and Mrs. Victor Paley also dined with us—afterwards the Montgomeries went to the Militia ball at Bury.

———

LETTER.

Barton, Bury St. Edmund's,
21st May, 1872.

My Dear Mary,

I was very happy to hear of your safe arrival *(your's in the dual)* in London, and that you had got so successfully through your French tour, without too much fatigue; I heard that *you* (in the singular) were looking particularly well, and I trust that Lyell finds himself rather the better than the worse for his expedition. It must have been very interesting, and I hope to hear a great many particulars about it when we meet in town, and especially to learn what observations Lyell was able to make in geology : what new ideas he gained, and what he now thinks of the Aurignac Cave. Also, I shall like to hear what were your impressions of the political state of France, so far as it is possible to 1872. judge of such matters in merely travelling through a country.

While you were away, I received vol. 2 of the new edition of the " Principles " for which many thanks : but I have not read much of it yet and indeed there does not seem to be much that is new since the last edition.

What a pity you were not at Naples when there was such a grand eruption of Vesuvius.

The weather seems to be improving a little now, but on the whole we have had a most wintry May : not worse, however, than many I remember in my younger days : and according to the newspapers it seems to have been just as bad in France. The flowers and foliage of the trees have been and are exceedingly beautiful in spite of the cold, and the colour of the grass is most brilliant. We have not for many years been at home all May, and it is a very beautiful month here.

You have heard of the great sorrow we have had in the loss of our charming and loveable young friend Mrs. Walrond. I have felt it, and do feel it deeply. And now we are anxious about our dear old friend Mr. Mills, who we have just heard is very ill, and he is in his 86th year.

With much love to Charles, believe me,

Your very affectionate brother,

CHARLES J. F. BUNBURY.

JOURNAL.

1872.   The Montgomeries went away in the middle of the day.   Arranged dried plants, and rambled about the park.   Poor little Peri, the paroquet, died rather suddenly.

———

Mr. James Gibson Craig arrived ; we walked with him through the arboretum, &c.   Mr. and Mrs. and Miss Hardcastle, Mrs. Vaughan Johnson, Lord John Hervey and young Arthur Hervey dined with us.*

———

We showed the library to Mr. Gibson Craig, and Fanny went with him into Bury.   In the evening we looked over my grandfather's drawings with him.

———

Mr. Gibson Craig went away in the afternoon.
A visit from Mrs. Hardcastle† and her sister Mrs. Vaughan Johnson:†

———

Fanny went to Mildenhall with Arthur and Miss Hamilton, returning to dinner.   .
Mr. Percy Smith dined with us, and Mrs. Smith came afterwards.

---

* Lord John Hervey and his cousin Arthur Hervey were in the militia, and therefore staying at the *Angel.*

† Both daughters of Lord Campbell, the late Chancellor.

The 28th anniversary of our happy wedding. God be thanked.

Began to make arrangements for leaving home.

Out of doors much of the day, enjoying the beauty of the weather and the flowers.

May 31st.

The weather beautiful and delightful since the 20th, especially for the last week or more; quite what May ought to be; most enjoyable. Barton has been in great beauty this month; the flowers and foliage as rich and beautiful as I have ever seen them, and the grass luxuriant. The Lilacs which were in exceeding beauty for some time after we last came home, are now quite past; the Laburnums, Judas tree, and Wistaria, almost so: but the Hawthorns, white and red, and the red Horse-chesnuts are still beautiful and so is the common Horse-chesnut in less sunny situations.

The Oaks and most of our trees, both native and exotic, are now in full foliage; the exceptions being the Ash among British trees, and among the exotic, principally the Catalpa, Deciduous Cypress, Gymnocladus and Gleditsia;—these two last, in particular, only now beginning to show their leaves.

LETTER.

1872.   My Dear Edward,

We propose to go to London on Monday next, the 3rd of June. There are many I shall be glad to see in town, but this place is in such beauty, that it is rather an effort to leave it. It is several years since we have spent the whole of May at Barton, and this has been a very beautiful one; a very great profusion of blossom, and the variety of colours of the young foliage very fine. Now it is rapidly passing into full summer, though the Hawthorns are not over yet, nor the Horse-chesnuts entirely.

We have been extremely quiet all this month, and I have been very well.

I am sure you must have been grieved to hear of the death of that most charming person Mrs. Walrond: it has been a great sorrow to us; it is years since anything has happened that has made me so unhappy.

Ever your affectionate brother,
CHARLES J. F. BUNBURY.

———

JOURNAL.

June 3rd.

We arrived in London. Mary Lyell came to see

us, looking very well, and giving a pretty good 1872.
account of her husband.

We are in great anxiety about my niece Emily,
who has fallen ill with small-pox : the last accounts
thank God, are better, and I trust the worst is
over.

———

June 5th.

We have a cheering account of our dear old
friend Mr. Mills, who had been dangerously ill, and
about whom we were painfully anxious. A serious
illness falling upon a man in his 80th year, and one
who has long been in rather delicate health, is
indeed an alarming thing. But, happily, there
seems now to be a good prospect of his recovery.

Dear Sarah Hervey arrived this day to stay
with us.

We have heard of the death of our neighbour Mr.
Milbank, who had long been seriously ill. I am
very sorry indeed for his wife Lady Susan, who is a
very excellent and very pleasing person.

———

June 6th.

John Moore told me the other day he hears from
Mr. Murray that Mrs. Somerville, now 93 or 94
years of age, is in wonderful preservation both of
body and mind. She can see to thread the smallest
needle without spectacles, and her needlework is
beautiful, and she watched the phenomena of the
eruption of Vesuvius with most eager attention and
interest.

1872.   We had a pleasant little luncheon party, at which, in particular, we had the great pleasure of welcoming dear Kate Hoare, just returned from her wedding tour on the Continent, looking brilliantly well and every way charming.

In addition to ourselves and Sarah Hervey, the party were Kate and her father, Minnie and Sarah Napier, Lady Rayleigh and Miss Kinloch.

<hr />

June 7th.

Charles and Mary Lyell came to luncheon with us. Charles looking tolerably well after his French tour. He appears to have enjoyed his tour, and to have been interested by various things that came under his observation, though the imperfection of his eye-sight must have been a sad hindrance to his enjoyment.

Lyells remarks concerning the great waterfall discovered of late years in British Guayana. It is situated at the head of an immense wall-sided ravine, which he believes to have been probably formed by *cutting back*, in the same manner as the ravine of Niagara : and he thinks it may serve to illustrate the mode of formation of the famous *canons* in California.

We (including of course Sarah) went with Kate Hoare to Fleet Street to see her home—the apartment in which she and her husband are now settled down for her married life. It is a set of very handsome and commodious rooms on the first floor of the house in which the Hoare's bank is carried

on ; Kate's drawing room a very pretty and well-
furnished room, quite removed from the roar
of the street. It was a great pleasure to see
dear sweet Kate thus settled in a cheerful home of
her own.

———— ————

June 10th.

Spent nearly an hour in the Royal Academy
Exhibition, taking a general view of the pictures.
Called on Emily Napier who was very pleasant.
Edward dined with us.

———— ————

June 11th.

William Napier, Edward and Cecil dined with us.

———— ————

June 13th.

Very fine. A drive in the Park with Fanny.
We dined with Lady Head. Met Mr. and Mrs.
Fuller Maitland, Boxall and a few others.

———— ————

June 14th.

We visited the Rhododendron show, (Antony
Waterer's Rhododendron tent) at the Horticultural
Gardens :—extremely beautiful. I have seen this
show in several previous years, but never enjoyed
its beauty more. I can hardly imagine that any
display of flowers, wild or cultivated can surpass it.

A beautiful day, quite warm. Emily Napier and
Susie came to lunch.

We heard at St. Peter's Eaton Square, a very
fine sermon from Dr. Barry, the Principal of King's
College, on the two great principles of Christian
morality—Truth (or Justice, or Righteousness) and
Love : how the one of these great classes of virtues
had been too exclusively cultivated and pushed to
an excess in some ages and nations, and the other
in others : and how both were reconciled and united
in the teaching of the Gospel and especially in the
life and doctrine of Our Lord.

———— ————

June 17th.

Splendid weather. Drove out with Fanny and
Sarah : we went to Kensington Gardens and walked
there. Our dinner party. Lady Mary Egerton and
one of her daughters, Leopold Powys and Lady
Mary, Lord Albert Seymour and his betrothed
Sarah, Lady Wilhelmina Legge and her brother
Heneage, Col. and Mrs. Lynedoch Gardiner,
George Napier, John Herbert and Admiral Spencer.
A very agreeable party. I sat by Lady Mary
Egerton, who is very pleasant, and for whom I have
a great esteem. In the evening a good many other
pleasant people. Lady Lilford, Minnie Powys (an
especially charming girl), Kate Hoare and her
husband, Mary and Katharine Lyell, Mrs. Hugh
Berners and one of her daughters; Mr. Eddis.

———— —

June 18th.

Our dinner party: Kate and Charles Hoare,

Lady Octavia Legge, the William Napiers, Mr. and Mrs. John Paley, Sarah Napier, Lord Albert, Minnie Powys, Mr. Medlycott, George Napier, Clement.

June 19th.

An afternoon party at the Charles Lyells : a good many pleasant people : Sir James Lacaita, Mr. Donne, John Carrick Moore and his sisters, Lady Rich, Fanny Mallet, Mr. Hutchings and many more. Had some talk with Mr. Donne. We dined at the William Napiers, only themselves and George and a Captain Hay.

June 20th.

Evelyn Bevan came to us, and we all (4) went to Colonel Gardiner's to hear Mrs. Scott Siddons read some scenes from Shakespeare and from Sheridan. She is really beautiful. I have rarely seen so lovely a face, and her voice and manner are very pleasing. The scenes she read were :—1st, some portions of "Much Ado About Nothing"—chiefly the dialogues between Benedick and Beatrice, which she gave with much spirit ; 2nd, the Sleep-Walking of Lady Macbeth : in this she was admirable ; 3rd, Henry the Fifth's Courtship of Katherine of France, given with a spirit and liveliness which made it very amusing, and which quite surprised me ; 4th, some of the scenes between Sir Peter and Lady Teazle— delightful. She read also a short poem by Whittier, " Maud Müller," which was rather pretty,

1872. and a long ballad of Shakespeare's Courtship, which
I thought rather dull. But her extreme beauty
recommended everything. I had not been used to
think highly of these Courtship scenes of Henry the
Fifth:—had thought them wofully inferior to the
rest of that play, and, indeed, hardly worthy of
Shakespeare. But Mrs. Scott Siddons read them
with such spirit and effect, brought out the contrast
so well, that she gave me quite a new idea of them.
Fanny and I dined with Mr. Cox and Lady Wood,
then Sarah joined us again and we all went to Lord
Hertford's, lastly the ladies went to a ball, and I
home.

June 23rd·

Our dearly loved Sarah Napier was married
yesterday, at St. Peter's, Eaton Square, to Lord
Albert Seymour. Most earnestly do I hope and
pray that this marriage may be productive of all,
and more than all the happiness which it seems to
promise so abundantly, and that they may be
preserved to a good old age to enjoy each other's
love and goodness. I am sure that nothing is
wanting in Sarah to make herself and others happy.
The wedding itself was a very pretty and pleasing
scene, and we were fortunate in a very fine day.
The service was admirably performed by Lord
Arthur Hervey and Dr. Blunt. All the arrangements were extremely good. Sarah looked even
more than usually lively, and both she and her
bridegroom appeared very happy. Sarah Hervey
was one of the bridesmaids, and by far the hand-

somest of them. The wedding breakfast was given
in our house, which we lent (at least the lower floors
of it) for the purpose. The whole celebration
passed off without a cloud: but before night there
was a sad and solemn contrast. Lady Napier died
in the course of that evening. She had been in a
very infirm and feeble state ever since she came to
London, and about a week ago she became so ill
that it was feared she might not live to see her
grandchild married, as she earnestly desired to do.
For her sake, the earliest possible day was fixed for
the wedding, and her health appeared to revive for
a time. Yesterday morning she appeared to be as
well as she had been for some time past. Sarah
saw her, talked with her and kissed her immediately
before the wedding, and she made some remarks on
her dress. In the afternoon, Minnie told her how
well all had gone off, and how the bride and bridegroom had started on their journey: she appeared
pleased and interested, and after a time said she
would rest: she fell asleep, and in that sleep she
passed away: nothing but her ceasing to breathe
marked the transition. There could not be an
easier or happier death. Lady Napier was of a
great age (in her 89th year, as her son John Herbert
tells me), and had long been in such a feeble state,
that her death must have been foreseen as
probable at any moment, yet it was rather sudden
and startling at last. She was an excellent woman,
of the best principles and warmest heart, always a
most warm, constant and cordial friend to me and
mine. She had lived to see the great wish of her

1872. heart fulfilled in the happy marriage of her darling
grandchild, and her death was truly a *Euthanasia*:
but it is natural that Minnie, who was devoted to
her mother, should feel it as a severe grief : she
loses at once the companionship of her daughter
and the occupation of taking care of her mother.

Monday, 24th June.

Miss Isabel Hervey and Miss Kinloch came to
luncheon.

Visited the International Exhibition, but was not
specially delighted. Edward and Clement dined
with us.

Tuesday, 25th June.

Went with Fanny to the National Gallery—afterwards to Hobart Place, and saw Minnie.

Mr. Walrond dined with us, quite alone—the first
time I have seen him since his great misfortune.

He was quite calm and composed, even gently
cheerful, and talked very agreeably on a variety
of subjects ; spoke of his lost wife with great
feeling, but with calmness, without appearance of
effort and without the least affectation.

He is a man for whom I have a great regard and
esteem.

June 26th, Wednesday.

Leopold and Lady Mary and Minnie Powys,

Octavia and Wilhelmina Legge, Isabel Hervey, and 1872.
Mr. Medlycott, came to luncheon ;—afterwards,
music.

———

June 27th, Thursday.

Went with Fanny to see Gustave Doré's great
picture of "Christ leaving the Prætorium;" now
exhibiting in Bond Street. It is really a grand and
noble picture, of very great power; a vast work,
indeed one of the largest oil paintings, I think, that
I have seen ; very powerful in all parts ; but the
figure of Christ has a dignity, a pathos, a beauty
and nobleness of expression, both in the face and
bearing, which surprised me. It gave me a much
higher opinion of Doré than I ever had before.

Afterwards we went to the Royal Academy. I do
not see any remarkable or striking pictures of
historical or poetical subjects ; but there are
beautiful landscapes, especially a delightful " Dewy
Eve," by *Vicat Cole*, a worthy counterpart to his
"Autumn Gold " of last year.

Beautiful sea-pieces by *Hook*, with his clear deep
green water, and rocks and sand and fish, recalling
the Cornwall and North Devon shores ; and others
in a different style, not less beautiful, by *Cooke*.
Ansdell's dogs and sheep — admirable. A lovely
little girl ("Helping the Gardener"), by *Eddis*. In
Riviere's "Daniel," the lions are admirable, the
Prophet insignificant. The picture which is most
talked about is *Millais'* portrait-group of the three
Miss Armstrongs ("Hearts are Trumps") : but the

1872. young ladies, I think, do not look as *real* as their
dresses, or as the flowers.

The Charles Hoares and Edward, dined with us.

———— ——-

Friday, 28th June.

We drank tea with the Charles Lyells, — met
Katharine also there—pleasant talk.

Clement dined with us.

———— ——

July 3rd.

We returned yesterday from spending two days
(and part of a third) very pleasantly with Frederick
Greys, at Lynwood, near Sunningdale, a very short
distance from one of the gates of Windsor Park.
Sarah Hervey was with us.   I have described, in a
former journal, the beautiful situation of Lynwood,
the fine and extensive view from it, and the
delightful country around.   Since we were there the
first time, much has been done in completing and
ornamenting the grounds, and there is an agreeable
and interesting mixture of wildness and cultivation,
beautiful garden flowers growing almost amidst the
heath and fern.   The house, as I think I mentioned
before, was built entirely according to Sir Frederick's plans and under his direction, and is, to my
thinking, the most perfectly well-designed and well-
arranged house that I know.

On Monday, the 1st, in the afternoon, Lady Grey
drove us out in the country, a very pretty and
pleasant way past Ascot race-course, and through
beautiful woodlands, which I think she called

Swinley Woods, and which, as I understand, form 1872.
part of Windsor Forest. The trees here were very
noble, and the fern under them tall and luxuriantly
beautiful. It was altogether an exceedingly agreeable drive.

On the 2nd the Greys took us to Windsor, a most
beautiful drive through the Great Park to the
Castle ; from the entrance of the Park to the Long
Walk it was the same way which they had taken us
in our former visit. The day was remarkably
fine, the scenery beautiful, and the drive most
enjoyable. One can hardly, I think, see anywhere
finer woodland scenery, nobler trees, or more
picturesque and beautiful groupings and dispositions
of them. The view from the high ground, when
one first catches sight of the Castle and the valley
of the Thames, is truly noble and impressive.

Having arrived at the Castle, the Greys showed
us the Chapel (Wolsey's), which is now restoring
and decorating in the most superb style as a
memorial to the Prince Consort. It is splendid
indeed. Here we took leave of our friends from
Lynwood.

Sarah Hervey went on with us to Slough, and
then, being joined by her youngest brother, started
off to join a water-party : we returned home.

We dined with the Youngs.

———

Thursday, 4th July.

While at Lynwood, we received the very sad
news of the death of poor Theresa Boileau. I feel

1872. deeply for her poor sister, who indeed cannot be
sufficiently pitied: they were devoted to each
other.

We went to luncheon with old Lady Lilford—a
very pleasant family party: then to tea with the
Charles Lyells, met Sir William Guise.

Our dinner party:—the Bishop of Bath and
Wells, Lady Mary Egerton, Sir Bartle and Miss
Frere, Sir John and Lady Hanmer, Philip and
Pamela Miles, Lady Head, Sir W. Boxall, Mr. and
Mrs. Longley, Mrs. Ellice, Mr. Clark and Captain
Wells—all very pleasant. In the evening a good
many other pleasant guests.

Friday, 5th July.

Charles and Mary Lyell and Mr. Church came to
luncheon: afterwards we went with Mr. Church
to the Deanery of St. Paul's, saw Mrs. Church, but
not the Dean: afterwards spent some time with
the Charles Hoares.

George and William dined with us.

July 6th.

Our dear Sarah Hervey left us, to my very
great regret, returning to Wells with her father,
her sisters Kate and Patience, and her brother-
in-law. I can hardly imagine a more charming
combination of graceful and kindly manners, lively
wit, refined and cultivated intellect, sweet temper,
warm heart, noble and elevated mind. I think her
very handsome, too.

Fanny went with Arthur to Haileybury and returned to dinner. I went to the British Museum. William dined with us.

———

Mrs. Ellice and her daughter came to luncheon. Fanny drove out with Minnie. We dined with the Maitland Wilsons.

———

We went to see the new Bethnal Green Museum. Going in our own carriage, it took us not quite an hour to go, and about the same to return, and I thought it worth even such a journey.

The hour-and-a-quarter we spent there, though as much as we could spare, was much too little for seeing the collection: indeed, one might spend many mornings there with satisfaction. Nearly all that is worth seeing, is the collection of pictures and of decorative art (porcelain, bronze, and the like) lately in the Hertford collection, and now sent for exhibition by Sir Richard Wallace. The pictures are a wonderfully rich collection: especially rich in the works of Reynolds and Gainsborough: in the French school of last century (Greuze, Boucher, Watteau, etc.), and in the Flemish and Dutch schools, but including also some fine examples of Italian art, and also of Spanish.

Kate Hoare came to luncheon. We dined with the Locke Kings. I sat by Lady Gore Browne

1872. (wife of the late Governor of New Zealand, and sister-in law of the Bishop of Ely), a clever, lively, agreeable woman. She told me much about New Zealand: particularly, that she had been very much struck by the resemblance in character between the Maoris and the Scottish Highlanders as described by Macaulay, so that she and her husband, when they read in New Zealand that part of Macaulay's History, felt that they might have been reading a description of the Maoris.

In dirt, in thriftlessness, and in disregard of truth (she said), as well as in courage, hospitality, generosity and devotion to their chiefs, the New Zealanders strongly resemble the picture drawn of the Highlanders. Their vices are such as are peculiarly annoying to the English settlers, and even the virtues of the two races are not calculated to reconcile them to each other.

Friday, 12th.

At luncheon at Lady Mary Egerton's. I had some conversation with Mr. Clements Markham. He said that great exertions are making by the South American governments, through the means of English and American companies, to explore and open up to navigation the many great rivers which flow into the Amazons. A plan for the steam navigation of the river Madeira has been matured ; and as the navigation of that river is impeded by 150 miles of almost continuous rapids, a railway has been planned to *turn* these impediments, and rails, carriages, workmen and surveyors have been sent out from England. He remarked how curiously sudden had been the transition, as to South America, from the most barbarous stage of road-making to the most advanced,—from mule-tracks to railways. He observed how many of the rivers which flow from the Peruvian and Bolivian Andes make a sudden bend in their course, almost at right angles, on entering the plains: their early course being nearly parallel to the main chain of the Andes: and he inferred from hence the existence of one (or more) subordinate exterior chains, like the Sewalik hills in India.

A railway is projected, guaranteed by the Peruvian Government, to cross the Andes and connect the coast of Peru with the feeders of the Amazons, and Mr. Markham does not think it impracticable. Mr. Markham observed, that the region of the Amazon river had been fortunate in having been explored by a number of remarkably able and learned travellers, beginning with La Cond-amine. He spoke highly of Mr. Spruce, the botanical traveller, and lamented that his continued ill-health prevented him from publishing any comprehensive account of his travels. Spruce (he said) had worked for several years in the forests of the hot regions, without suffering materially in health: but what broke him down was a rapid journey up to the cold ridges of the Andes of Quito; the valleys in that country are so deep, and the mountains so steep, that you ascend in one day from extreme tropical heat to a very cold climate.

Spruce was attacked with fever (I suppose from

1872. catching cold), and it ended in something like paralysis.

We went to the Zoological Gardens,—then to see Katharine—then to a garden party at Mrs. Marlay's.*

Minnie and Edward dined with us.

July 13th.

We visited my old Cape friend, Sir John Bell, a very fine old man, now (I believe) in his 91st year, —extremely deaf, but otherwise in wonderful preservation, his sight good, his mind vigorous and perfectly clear, his talk humorous and animated as of old.

Fanny went out to an evening party at Madame de Bunsen's

Monday 15th July.

Our dinner party:— the Brampton Gurdons, Sir Edward Ryan, the Wilsons, Captain and Mrs. and Miss Berners, Admiral Spencer, Mr. Hutchings, General and Mrs. Eyre.

July 16th.

We drove to Sydenham, to the Crystal Palace, and saw the new Aquarium, which is very fine of its kind, very curious and well-worth seeing. It is curious and entertaining to see, in the large tanks of salt water, a great variety of fish, crustaceans and

---

* Mother of Lady John Manners, daughter of Lady Charlemont. She was living on the North side of the Regent's Park.

other marine creatures, moving freely about in their natural ways, with plenty of room and quite at their ease. In one tank, in particular, there were several large lobsters (not the common kind, but more spiny and of different colours), and their manner of walking, of twirling their feelers, and occasionally combing themselves with their small claws, was very grotesque.

In another were a number of Labridac (Wrasses), some of them of beautiful colours,—fishes I had often read of, but neyer before seen alive: and without seeing them alive, one can have no idea of the beauty of their colouring. In another, several kinds of flat fish (Pleuronectidac), such as soles, flounders, plaice and others—creatures which would be thought marvellously strange and grotesque if one were not used to them.

We dined with Lord and Lady Hertford. After dinner Lord Hertford showed me a few interesting pictures, which he has in his dining room:—in particular, a beautiful portrait, known as "The Lady in Blue," by *Sir Joshua Reynolds*, (this I saw in the Royal Academy Loan Exhibition, in February); Mrs. Robinson (Perdita), also by *Sir Joshua*, and beautiful; the portrait of Horace Walpole (standing and leaning on his elbow), which I knew by the engraving:—and a portrait of Marshal Conway.

Met Lady Frederick Kerr, Mr. Blunt and many whom we did not know.

Friday, 19th July.

A beautiful day, very warm. Visited the South
Kensington Museum. Minnie, Edward and Clement
dined with us.

LETTER.

Bristol Hotel,
Brighton, July 22nd, 1872.

My Dear Joanna,

Fanny begs me to thank you very much
for a very agreeable letter which she received from
you the other day from Chaumont. She is so excessively busy with a variety of matters, and
especially with the care of Arthur MacMurdo and
the arrangements for sending him (next year) to
school, that she has no time for writing a letter
"of facts" (so she says, and I suppose still less
for one of fiction;—and so she has asked me to
write. But I do not know how far your knowledge
of our facts has been brought down. To begin with
the present :—we came down here on Saturday, the
day before yesterday, but our excursion hitherto has
not been very prosperous : the day we arrived was
overpoweringly hot, and so was yesterday. Fanny
has been almost ill with the heat, and little Arthur,
who was delighted with the sea and eager to bathe,
was made quite ill by the effect of the sun on his
bare head, and rather alarmed us yesterday ; this
morning, however, he seems pretty well. This
morning we have had a violent thunder-storm, which

is still going on. We propose to return to town 1872. either to-morrow or Wednesday, and hope to get down to Barton next Saturday at the latest; and very happy we shall both be to find ourselves there. Last Tuesday, the 16th, we drove to Sydenham (*we* including little Arthur) to the Crystal Palace, and saw the new aquarium, which is a very fine one and very curious. In the evening of the same day we dined at Lord Hertford's, and I had a good deal of talk with an agreeable clergyman, a Mr. Blunt, who is going to marry one of Lord Hertford's daughters.

On the 18th (Fanny's birthday), dear Kate Hoare came to luncheon with us, and we took leave of her: very sorry I was to do so; but she will be within much easier reach of us (being in London) than the family at Wells, and I hope to see her again in the autumn. She and her husband look the very personification of good humour, easy temper, and happiness. (This is a parenthesis).

Afterwards we drove out to Sir Bartle Frere's, on Wimbledon Common, and saw something of the camp and of the rifle-shooting for the Elcho Shield —a pretty sight (I mean the camp).

I suppose Fanny has told you that on the 15th, this day week (it was very cold by the bye), we visited Katharine, and were very glad to see the Pertzes all looking well. They are now I hope all enjoying themselves at Barmouth. I am very much interested by Grotes' "History of Greece," in which I have now read as far as the Battle of Delium, in the 8th year of the Peloponnesian War :

1872. but in London, especially of late, the regularity
of my reading has been a good deal interrupted. I
hope to read well at Barton. I hope you will much
enjoy your summer at Chaumont, and meet with
nothing but what is pleasant, though I can well
understand that it must be painful to you and
Susan to be separated.* *Her* we hope to see at
Barton before very long. Dear Minnie will either
go down with us to Barton or join us there soon
after.

Believe me ever your very affectionate brother,

C. J. F. BUNBURY.

I am *very* sorry we have no chance of seeing you
at Barton this year.

JOURNAL.

July 29th.

We returned home to Barton on Saturday last,
the 27th, dear Minnie Napier with us. The weather
intensely hot, as it had been for a week past, and
as it still continues. All safe and well at our dear
home, thanks be to God. This place is in great
beauty, for the very heavy rains which have fallen
from time to time, joined to the heat, have been
extremely favourable to growth, and I have seldom
seen vegetation here more luxuriant. Grass brilliantly green, foliage very rich, and flowers in great
beauty. The Catalpa tree is in superb blossom ;
the Pomegranate more luxuriant and rich in foliage

* Susan went to Paris for a short time.

than I have ever seen it ; two beautiful Clematises 1872.
Jackmannii and lanuginosa) in flower. Tilia alba
in full flower, spreading its honied fragrance far and
wide. Æsculus (Pavia) Indica, past flowering;
Æsculus macrostachya just beginning. Magnolia
grandiflora showing many flower buds. An abundance of cones on one of the Lebanon Cedars
on the lawn ; but on one only.

Settling ourselves at home.

Tuesday, 30th.

We dined with the Rodwells at Ampton—met
Judge Keating, Lady Cullum, the Wilsons, the
Chapmans.

====

LETTER.

Barton,
July 30th, 1872.

My Dear Henry,

Many thanks for your letter. We arrived
here safe and well on Saturday, and were thankful
to find all well at home ; we had a very hot
journey and rather a troublesome one, as the train
was very much crowded. Dear Minnie came with
us, and altogether we were a largish party : four
maîtres (including Arthur), seven servants and four
parrakeets.

Very heartily glad I was to get out of London,
and still more to find myself here ; such a relief

1872. after that broiling and noisy town.   Sunday was
indeed very hot, even here, and so was yesterday
morning, but in the afternoon the weather changed,
and to-day we have heavy rain.   This place is
in great beauty for the abundance of rain which has
fallen during this month, combined with the heat,
has exceedingly promoted vegetation.   The grass is
as bright as in May, the foliage very rich and the
flowers beautiful.   The Catalpa in the arboretum is
in great beauty, the Pomegranate more luxuriant
than I have ever seen it before, and there is a great
show of buds on the Magnolia grandiflora.   The
flowering of Pavia Indica is past.   There is a *lot* of
cones on one of the Cedars on the lawn, but on only
one,—the one *beyond* Cissy's.

A striking effect of the combination of heat and
wet is the appearance of a great crop of large
toadstools (Boletus luridus), which usually do not
appear till the autumn, under the trees near the
arboretum.   Altogether it is very *fungus-ish* weather,
and I see that Scott is uneasy about its effect, even
on the wheat : the potatoes of course will suffer.
If, however, we are blessed with fine dry weather
for the harvest, the corn-crops seem likely to be
good; in this parish the wheat harvest is expected
to begin this week.   There has been no strike here
as yet, and Scott tells me he believes that the
principle farmers have been able to secure a
sufficient number of labourers for the harvest, though
at rather higher rates than usual.

Fanny suffered a little from the heat, which
prevented her from sleeping well, but to-day, as it is

cooler, she seems better, and she looks very well. 1872.
She works as usual incessantly, especially for
Arthur's interests, her heart being very much set on
having him prepared to enter at Haileybury next
spring.

I am very well, and was so during the hottest of
the hot weather; I wish we may not have too great
a change : I am rather like a snake in this respect.

Dear Minnie seems pretty well.

With much love to Cissy, Emmy and the boys,

Believe me ever,

Your affectionate brother,

CHARLES J. F. BUNBURY.

We had a visit from Lady Cullum yesterday.
To-morrow I go on the Grand Jury.

JOURNAL.

Wednesday, 31st July.

The Assizes at Bury :—

I was foreman of the grand jury—a light calendar
—grand jury discharged before 2 p.m.

Dined with the Judges, Byles and Keating,—both
very pleasant.

LETTER.

Barton,
August 5th, 1872.

My Dear Katharine,

I will now send you a *supplement* to my
former letter, as I certainly have now no heat to

1872. complain of. There has been indeed a lamentable
change in the weather ; last Monday, this day week,
was the last of the heat, and ever since we have had
nothing but rain storms, excessive damp and raw
chilliness ! unpleasant enough as to our personal
sensations, but most deplorable as affecting the
harvest, which ought to be now in full course.
There is great reason to fear that the wheat will be
attacked by mildew and very seriously injured, in
consequence of the excessive damp ; indeed, unless
very fine hot dry weather sets in very soon (of which
there is no apparent prospect), I fear the crop will
certainly be very deficient. And then there is the
foot-and-mouth disease very prevalent among cattle,
and great fear of the importation of the cattle
plague ; so that altogether our prospects are not
pleasant. But we must only hope and trust that
all may be for the best. I have not for many years
seen this place looking so thoroughly damp, so
soaked and sodden as it is now. We shall soon be
all over blue mould as they say. It is a fine season
for Fungi, and they are coming up everywhere
in prodigious abundance, just as they usually do
in the middle of autumn ; particular abundance of a
great Boletus (Boletus luridus) which I do not
usually see before October. The weather is good
also for the grass, the foliage, and the ferns, but the
flowers have been much knocked about by the
heavy rains. In the conservatory and the hothouses
however, we have some beautiful flowers: Allemanda
Schottii, very fine, with plenty of its great yellow
flowers, which you may probably have seen at the

flower shows in London ; several plants of Lilium 1872.
auratum in fine flower ; Vinca rosea covered with
blossoms ; Passiflora racemosa, and others.

I am very glad that you have been enjoying
Barmouth so much, and so delighted (as I do not
wonder you are) with the beauty of the country. I
hope the rain has not spoiled your enjoyment.
Here, in spite of the bad weather, we are enjoying
the quiet of home and our various occupations ; and
it is a great pleasure to have dear Minnie's society.
Fanny is pretty well, and excessively busy. Arthur
in great force, and the birds *(nine* in number), very
flourishing. The little " Waved Parrakeets," or
Budgerees, are the prettiest little creatures possible.

I heartily hope that the Kew affair may be
settled so as to make Dr. Hooker comfortable.
The only entirely satisfactory conclusion, indeed,
would be Mr. Ayrton's retirement ; but I am afraid
he has no idea of the Christian virtue of resignation.

Ever your affectionate brother,

CHARLES J. F. BUNBURY.

JOURNAL.

August 7th.

Fanny received a letter from Rose Kingsley, from
Eversley ; I am exceedingly glad to hear of her safe
return home, after her very adventurous (and no
doubt also very interesting) journey in North
America and Mexico.

She gives a good account of her father's state of

1872. health, but, I am sorry to say, not so good of her mother.

Read, in *The Times*, the copious extracts from Dr. Livingstone's despatches :—curious, if not entirely satisfactory. His accounts of what he has discovered concerning the water-shed of Central Africa, are not very clear—at least they require maps and farther illustrations : probably they will before long be made more intelligible by discussion at the Geographical Society. He gives a striking description of some of the difficulties and dangers he met with—difficulties, not owing (according to our old received ideas of Africa), to drought and sterility, arid deserts and want of water, but to excess of water, to swamps and bottomless mud, innumerable streams, and rank luxuriance of vegetation.

August 10th.

Weather very bad almost ever since the end of last month—excessive rains, few days without rain, and hardly one really warm. A sad prospect for the harvest.

August 12th.

Received a charming letter from Sarah Hervey, expatiating with particular delight on their visit to Clovelly, with which she is perfectly enraptured, and I do not wonder. It is indeed a very beautiful and a very singular place.

Received an interesting letter from Charles Kingsley, from Eversley.

Speaking of the weather (about which I have been rather *croaking*), he says:—" The glass has risen, "steadily and high, here, on a blustering S. W. "wind—the very best sign for fine weather. The "wind has now fallen up to N. N. W., with clear "dry air and sun, and there is every prospect of "good harvest weather. I have trusted all through "that the excess of rain during the last six months "would be followed by fine August weather: the "continuous thunder made me more hopeful still, "as it seemed likely to clear the air permanently of "the great weight of vapour which accumulated "there from the sunbeams of spring evaporating the "great winter rainfall off an earth not chilled by "winter frost."

LETTER.

Barton,
August 15th, 1872.

My Dear Lyell,

I have had a letter from Charles Kingsley, containing a suggestion on which he wishes for your judgment. He says:—" As for Hooker's case, you "cannot be more indignant than I am. This is a "good opportunity for Science in general to show its "strength, and I cannot but think that if there is "any fear of Hooker's resigning, we might get up a

1872. "representation signed by all the members of local
"scientific societies, which would give the Ministry
"some idea of our numbers and importance. I
"could get from 2-300 signatures in Chester alone.
"Think of this, and if it seems good, will you ask
"Sir C. Lyell what he thinks of it?"

So far Kingsley. The expediency of such a step
as he suggests, must depend mainly, I think, upon
what Hooker feels as to his position. The last
"Treasury Minute," which closes the published
correspondence, seemed tolerably satisfactory, but
nothing could be more thoroughly *nasty* than Mr.
Ayrton's speech in the debate on the subject, and I
was very much disatisfied with Mr. Gladstone's. I
am afraid, after all that has past, that Hooker is not
likely to feel thoroughly comfortable with Mr.
Ayrton for his official superior ; but unless he
seriously contemplates resigning, I am not sure that
any further stirring in the matter at present may not
do him more harm than good.

Kingsley writes, that Rose has come back far
stouter and stronger than when she went, and laden
with curiosities of all sorts. Her journals must be
very interesting ; I hope she will publish.

I hope you enjoyed your stay at Barmouth, and
that it agreed with you.

With much love from both of us to Mary and
yourself,

Believe me ever,

Your affectionate brother,

CHARLES J. F. BUNBURY.

Saturday 17th.

Arthur's cricket party. Mrs. and Miss Wilson and Mrs. and Miss Praed came with a number of boys and spent the afternoon with us. Lady Louisa Legge went away in the afternoon.

Monday, 19th August.

A great change of weather for the better since my note of the 10th. The whole of last week since Sunday the 11th, was fine, aud the last three days have been beautiful, so that the harvest has made rapid and most satisfactory progress, and the farmers are joyous. A great quantity both of wheat and barley has been cut and stacked in good condition in the course of the week.

A beautiful day. We arranged the newly bound books. Had a pleasant walk with Minnie.

Tuesday, 20th August.

Examined the Ferns of Allamanda Schottii (?) I am not quite sure of the species) which has flowered in our hot-house : and made a note of it.

George and Leonora Pertz and their two daughters arrived.

Wednesday, 21st August.

A beautiful day. Had a pleasant drive with Minnie to Thurston and Rougham.

Thursday, 22nd August.

Mr. Medlycott arrived. I showed him the gardens.

Friday, 23rd August.

Lady Cullum, Mr. Abraham and his daughter, Mr. Wrattislaw, and Mr. Littlewood dined with us.

- - - -

Saturday, 24th August.

Splendid weather.   Mr. Medlycott went away. Fanny went with Leonora, Annie and Arthur to the Praeds at Ousden, and returned late.

I had a pleasant walk with Minnie.   The Louis Mallets arrived late.

- - - -

August 26th.

The wheat harvest in this parish is, I believe, now finished, and the crop has been got up in good condition, a good deal of the barley is still out.

- - - -

August 27th.

Dear Minnie left us to return to London, I was sorry to part with her: she has been a most pleasant companion, I look upon her quite as a sister.   The Pertzes also went away.   Walked with Louis Mallet.

═ ═ ═

LETTER.

Barton,
August 27th, 72.

My Dear Mary,

Many thanks for your letter of the 19th.   I communicated to Mrs. Kingsley as much of it as

related to Dr. Hooker, but have not heard anything
more.

Minnie Napier has gone to-day with the Pertzes
—gone to London. I am very sorry to part with
her.

I read Galton on "Hereditary Genius" some
time ago, and thought it very interesting, containing
a great store of well-arranged facts, and showing
immense industry, but I was not convinced of the
truth of his theory. There was what I thought a
very good article upon it in the *Edinburgh Review* (I
cannot at this moment refer to the date)—showing
some of his weak points:—in particular, that he
has much underrated the influence of education, in
the widest sense (meaning the aggregate of all the
external circumstances which act upon us in early
life), the result of which is with great difficulty
distinguished from innate qualities. Old James
Mill, I remember, affirmed that *all* the differences
between one man and another were owing to
education in this wide sense. This I do not believe,
but neither am I satisfied of the entire soundness of
Galton's doctrine. Galton's paper on Blood
Relationships, in the Royal Society's Proceedings,
is very hard to understand.

I see, from the discussions at Brighton, which I
have been reading in *Nature*, that the geographers
seem a good deal puzzled by Livingstone's discoveries, and at a loss to determine how far they
throw any real light at all on the upper course of
the Nile. I certainly thought his letters as
published very far from clear as to geography: but,

1872. indeed, Mr. Stanley's paper also seemed rather unintelligible without the help of maps, such as I suppose were before the eyes of the section while he lectured. I am very glad that Livingstone has been found safe and well, but it seems that we must yet wait for much more information about Central Africa.

We have beautiful weather here—beginning to feel decidedly autumnal, and the trees and flowers are in beauty, the grass uncommonly rich and brilliant.

I hope you will come and see us on your way back to London. At present, we have (besides Mrs Byrne) the Louis Mallets with us, who are very pleasant. Mallet is very agreeable : though an official, he fully sympathizes with us about Hooker.

With much love to Lyell, and kindest remembrances to the Miss Lyells,

Believe me ever,

Your affectionate brother,

CHARLES J. F. BUNBURY.

STUDIES.

Chapter 50 of Grote's "Greece" includes some of 1872.
the most terrible acts of vengeance and cruelty
which stain the annals of Greece. First, the revolt
of Mitylené (one of the allied cities) from Athens, its
re-conquest, the atrocious decree (that the whole
male population of military age should be put to
death) proposed by Cleon and carried by his
influence in the Athenian Assembly, the speedy
repentance of the Athenians, and the dispatching
(only just in time) of a milder decree :—not very
mild either, for it ordained the execution of all who
were considered as active in the revolt, to the
number of upwards of a thousand. Secondly, the
conclusion of the siege of Plataea, its surrender, and
the cruel massacre of its garrison by the Lacedæmonians. Thirdly, the civil war in Corcyra, and
the barbarous massacre of the defeated party,
certainly very analogous to some of the proceedings
of the Jacobins in the French Revolution. In
chapter 51, we have the first appearance of Nicias
(or Nikias) on the scene, and the victories of
Demosthenes, near Ambracia. Chapter 52, com-
prising the 7th year of the war, is chiefly occupied
by the memorable success (exploit) of the
Athenians, B. C. 424, in the island of Sphacteria
(at the entrance of the bay of Navarino) where they
made a considerable number of the best soldiers
(hoplitae) of Sparta lay down their arms and

1872. surrender as prisoners of war. Such a mortification to the military pride of Sparta was unexampled, and seems to have created an immense sensation throughout Greece; in the Athenians, as was natural, it excited extravagant hopes of triumph.

All these operations are very well told by Grote, with great clearness and spirit.

It was in this campaign that the great Brasidas appears to have first distinguished himself.*

The number of victims at Plataea was much less than in the case of Mitylené (because the greater number of the defenders had previously made their escape from the town), but the deed was, perhaps, even more atrocious, at any rate more inexcusable.

Grote is always very partial to Cleon, but he does seem to be right in contending that Cleon gave good advice with respect to the Sphacterian expedition, and that his counsels contributed in an important degree to the success. The military merit, however (as Grote admits), was entirely due to Demosthenes.

Chapter 53,—the 8th year of the Peloponnessian War:—active and important military operations, the result of which was very unfavourable to the Athenians, and quite counteracted the effect of their victory at Sphacteria. They were totally defeated with great loss, by the Thebans and allied Boeotians in the great battle of Delium—very well told by Grote. Soon after this came the series of brilliant successes of Brasidas in Macedonia and

---

* Not quite correct; his first exploit was the defence of Methoné (Modon), against the Athenians, in the first year of the war.

Thrace, and especially his capture of Amphipolis, a 1872. most severe and mortifying blow to the Athenians.

Brasidas was a really great man — great in character as well as in military skill and ability, one of the greatest men that Sparta ever produced. It was in this, the 8th year of the war, that the Spartan rulers (the Ephori), perpetrated one of the most extraordinary acts of treacherous cruelty recorded in history:—selecting under pretence of honour and reward, 2,000 of the bravest and most ambitious and enterprising of the Helots, and then secretly destroying every one of them,—so secretly that, Thucydides says, it was never known what became of any of them.

The principal event in the 54th chapter is the battle (or skirmish) of Amphipolis: but a slight affair as to fighting, yet very important, and indeed very peculiar, from the fact of the commanders of *both* armies, Brasidas and Cleon being killed. The Athenian forces were disgracefully beaten; yet, as Grote justly says, the death of Brasidas converted their defeat into a substantial victory, and a very material one, for there was no man on the Lacedæmonian side at all qualified to take his place. It happened, moreover, that the two leaders thus slain were the two greatest opponents of peace in their respective countries, and thus their deaths opened the way to a temporary accomodation.

The volume ends with the conclusion of the so-called Peace of Nicias, in the spring of 421, B. C., after the war had lasted ten full years. It was professedly intended to last 50 years.

1872. Chapter 55, from the Peace of Nicias, March 421, B. C., to the Olympic festival of the 90th Olympiad, July, 420. Not a very interesting chapter, except for the first appearance of Alcibiades on the scene.

The character of this extraordinary man is very vigorously drawn by Grote, and I daresay justly. He had some resemblance, it strikes me, to the Duke of Wharton, as drawn by Pope and by Macaulay : but even Wharton, perhaps, hardly came up to Alcibiades, either in brilliancy of talents and accomplishments, or in impudent immorality both public and private.

Chapter 56, the battle of Mantineia, fought in June, 418, B. C., extremely well and clearly told. This was of course not the most famous battle of Mantineia, that which was made memorable by the death of Epaminondas, in 362, B.C. But this was also a great and important battle — important, especially as it completely restored the military reputation and ascendancy of Sparta. It was fought between two leagues or alliances of states : —on the one side the Argeians (or Argives), Athenians and Mantineians : on the other, the Lacedæmonians, Tegeans and some lesser contingents from Arcadia.

One incident of the battle reminds one of Prince Rupert, at Edgehill and Naseby. The Mantineians and a select portion of the Argive troops, who formed the right wing of their army, totally routed the left wing of the Lacedæmonians, and pursued it to the baggage-waggons, which they took; but they neglected to take the Spartan centre in flank, as

they might have done, or to support their own : and 1872.
while they were wasting their time in the pursuit,
the Spartan centre and right carried everything
before them, and gained a complete victory. The
narrative of this battle gives us some information on
the tatics of Greek armies in the 5th century, B. C.
It concludes with the atrocious massacre of the
male population of Melos,* by the·Athenians : an
act which even Grote denounces as "one of the
grossest and most inexcusable pieces of cruelty
combined with injustice which Grecian history
presents to us."

The special atrocity, as he observes, consisted in
this : that the Melians had not been at war with
Athens, nor had given her any provocation : they
had never been subject to her, and therefore had
not revolted. They only wished to keep their
neutrality, whereas the Athenians were determined
that all the islands should be *their* subjects. The
Peace of Nicias lasted in reality but a very short
time : it was not formally abrogated till after the
Melian massacre, but the Spartans made no attempt
to carry out its provisions on their part : both
parties intrigued and tried new combinations, with
the view of weakening each other : military operations were undertaken, and even battles fought
during the nominal continuance of the peace. The
Athenians endeavoured to set up the Argeian power
as a counterpoise within Peloponnesus to that of the

---

* The island now called Milo. The Athenians decreed to kill all the males of
  military age, and to sell the women and children for slaves.

1872. Spartans, but the battle of Mantineia defeated this scheme.

Chapter 57 of Grote; the preliminaries of the Athenian expedition against Syracuse; arrival of envoys from Egesta and Leontini to beg the assistance of Athens; debate in the Athenian assembly and antagonist speeches of Nicias and Alcibiades, the latter most mischievous : — decision of the Athenians to send a great expedition to Sicily.

Chapter 58: the strange incidents of the mutilation of " The Hermae," and the profanation of the Mysteries. Grote dwells at considerable length on this singular chapter of history, and labours hard to make out a tolerable case for the Athenian *Demos.* I hardly know why it is that we find it so difficult to understand,—almost to believe,—the genuineness of that terrible outburst of religious fanaticism and bigotry among the Athenians; an explosion of fury originating in superstitious terror. We are apt, almost unconsciously, to fancy that it could not really have arisen from real religious feeling. Perhaps it is that we are so much used to associate Grecian religion with the beautiful forms of art, that we do not easily bring ourselves to think of it in its more serious and stern aspects. Perhaps we are unconsciously influenced by a *Gibbonism* which leads us to look at ancient religion rather from the philosopher's than the people's point of view.

Certainly it is most difficult to understand that feeling or belief which, according to Grote, would have been most natural to an ancient Greek :—

namely that there was a clear connexion between 1872.
acts of audacious impiety and treasonable designs
against the State.

The analogy between the state of the public
mind at Athens on the occasion in question, and
that of the English people at the time of the
"Popish Plot" is very striking, and the parallel
is, I think, closer than Grote admits. The panic of
the Popish Plot was not, as he says, entirely
baseless. The murder of Sir Edmondsbury Godfrey,
whoever might be the authors of it, was a real fact,
as real as the mutilation of the Hermae; and it was
enough, under the circumstances of the time, to
give the first occasion for that frenzied terror, of
which Oats and his accomplices took such villainous
advantage. Considering that such a man as Lord
Russell seriously believed in the Popish Plot, we
need not wonder at the credulity of the Athenians.

It is very hard, in spite of all that Grote says, to
understand what could have been the object or
motive of the *mutilators*. If they desired to make a
revolution and upset the existing government by
force, one would have thought that such a public
outrage, which must rouse and alarm the authorities
would have been more likely to defeat than to
advance their object.

Chapter 58 of Grote :—The arrival of the Athenian armament in Sicily; the flight of Alcibiades to
Peloponnesus (in consequence of the machinations
of his enemies at Athens), and his revolt against his
own country;—preliminaries to the Athenian siege
of Syracuse. Nicias certainly seems to have been a
wretched general.

JOURNAL.

1872.   A beautiful day.

We inspected the Ferns, and Fanny began a catalogue of them.

_____

Walked round my farm.   The Abrahams, the Wilsons, and the Chapmans dined with us.

_____

Louis and Fanny Mallet went away; they are very pleasant.   Louis Mallet is now a member of the Council for India; a very good man, a thoughtful, earnest man, and of great ability.   His wife is very handsome and very pleasing.

Louis Mallet said to me, the two men in the Indian Council who are most congenial to him, and whom he considers as the ablest of all, are Sir Bartle Frere and Sir Henry Maine.   He does not think that our Indian dominion is in any *immediate* danger, but it can never be considered as really and permanently secure.   The chief difficulty and danger at present lies in the finances.   The public debt of British India, incurred by the Government chiefly for the making of public works amounts as I understand him, to about 200 millions sterling; and this though *absolutely* so much less than our national debt, bears a larger proportion to the total income of the country.   Of this debt, nearly the whole is held by people of Great Britain : therefore, if we should lose India, the whole of the loss on account of the debt would fall on this country, while the Indians would be relieved from the burden of it, and at the same time would retain all the advantages of the public works, on account of which it was incurred. In the meantime heavy taxes are required to pay the interest of the debt, and are of course very unpopular.

It is exceedingly difficult to find taxes suitable to the condition and habits of the people.

The income tax falls on only a very small proportion of the population; those that escape it by reason of their poverty being many times more numerous in proportion than in any West-European nation.

One of the difficulties (Mallet said) which a European has in understanding India, is that of believing the great and wide spread *poverty* of the people.. (The same thing, I remember, is remarked by Macaulay). There has so long prevailed in Europe, a popular, traditional, almost mythical notion of the wealth of India, that people do not readily bring themselves to believe what is the truth, that the very great majority of the population are much poorer than the lower classes of any country of Western Europe. An immense proportion of the people of India (he says) have not more on an average than £5 sterling per annum.

Louis Mallet told us, that posts in the *uncovenanted* civil service of India have, of late years been thrown open to unlimited competition, to

1872. natives as well as Europeans, and that the natives
are found to *beat* the Europeans *hollow*,—to surpass
them by far in the examinations. The authorities
are therefore placed in this dilemma—they must
either have a great portion of the offices filled by
natives, or they must give up their dearly beloved
principle of competition. A plan proposed for
collecting information relative to useful natural
products of India,—a commission to collect such
information by circulating forms of inquiry in all the
several districts. This plan vehemently opposed in
the council of India, but warmly advocated by Sir
Bartle Frere, and favoured by Duke of Argyll:
ultimately recommended to Lord Northbrook (L.
Mallet).

Attended the Special Sessions at Bury —
discussion of division of County into Polling
Districts, and on County Licensing Commission.

Thursday, 5th.

Beautiful weather.

Long discussion with Fanny and Scott about
labourers' wages and London houses—satisfactory.

Saturday, 7th September.

Mrs. Byrne went away—also Arthur's governess.
Susan Horner arrived late, looking very well—also
George and John Freeman.

Monday, 9th.

Received a very interesting letter from Charles

Lyell on the subject of Galton's "Hereditary 1872. Genius," and specially on the extraordinary development of intellect in Attica, in the century, from 530 to 430, B.C.

Had a pleasant walk with Susan.

Wednesday, 11th.

Engaged at Bury for nearly three hours on the Committee for re-arranging the Polling Districts under the Ballot Act. Maitland Wilson in the chair.

Thursday, 12th.

Showed some of the minerals to the Freeman boys.

Friday, 13th.

A beautiful day.

The Freeman boys went away. Walked to the Gravel Pits, and found Linaria minor.

Saturday, 14th.

Drove with Fanny and Susan to see Mrs. Rickards, and then round by Nether Hall and Pakenham.

Sunday, 15th September.

A very lamentable event happened here yesterday. Mr. Paine, one of the most sensible, intelligent and

1872. useful men in the parish, a good and valuable man in every way, was riding an ill-tempered and unmanageable horse, against which he had been cautioned: it took fright, shyed sideways against some railings at the side of the road, broke down the railing, fell with its rider, rolled over him, and set its feet on him in rising. Mr. Paine lingered a few hours, and died towards 7 o'clock that evening.

I feel it as a real misfortune, both to myself and the parish. Except Scott, I do not know any other man whose death would have been so great a loss to this place.

––––––

September 17th.

A large and handsome old oak in the park here, near the west lodge, broke asunder last night or this morning, though there was little wind: more than half of its head and a great portion of the trunk broke off and fell down, the tree being in fact torn into two. The inside was extensively decayed, the process of decay having evidently begun at the top and worked downwards, and at the bottom of the decayed part, far down in the trunk, was found an owl's nest, with a live owl on it.

══════

LETTER.

Barton,
September 21st, 1872.

My Dear Henry,

What you say about yourself makes me very anxious and unhappy. I hope and pray that the state of the case may not be so bad as you 1872. suppose.

I am afraid the soft relaxing air of the West always affects your constitution. I am very glad you are thinking of visiting Adair in the Highlands, and I hope the bracing air will do you much good. I shall be very anxious to hear again of your health. Here the weather is excessively cold (with a north wind), and has been so for the last four days ; but we are both tolerably well, though the cold interferes with our comfort.

Your letter crossed mine in which I told you of the loss Barton has suffered in the death of poor Mr. Paine. His funeral was yesterday, and was very well attended, I hear ; but unfortunately I was not able to be there, as we did not return in time from Riddlesworth, to which we had been engaged for some time.

With much love to Cissy and Emmy,
Believe me ever,
Your affectionate brother,
CHARLES J. F. BUNBURY.

JOURNAL.

Monday, 23rd.

Katharine and Rosamond Lyell arrived.

Tuesday, 24th.

Scott gave me a very good report of Mildenhall Rent Audit.

Charles and Mary arrived

1872. Looked over some of my dried plants with Katharine.

Very stormy and wet.

Looked over some more dried plants with Katharine. Walked with Charles Lyell. Fanny, Mary and Susan went to Mildenhall and returned to dinner. Herbert arrived.

Katharine showed me a set of drawings of Natal plants.

Helped Susan to correct her MS. relating to the Florence Museum.

Mr. Hutchings arrived.

The Charles Lyells, Katharine and Rosamond and Susan went away. I was very sorry to part with them. Susan has been with us since the 7th, and has been exceedingly pleasant the whole time: she is in very good looks, very young for her age, in excellent spirits, and full of animation and activity, with great talents for amusement as well as for more serious objects, with a great deal of knowledge also ; in short, remarkably agreeable. She has been busy during her stay here, correcting the proof-sheets 1872. of an elaborate work on Florence, on which she and Joanna have been engaged for some years, and which is to be published by Strahan.

Charles Lyell is, I grieve to say, very infirm in bodily health, and moreover very nearly blind, so that he requires constant care.

His doctors insist particularly on his avoiding all fatigue, either of body or mind; and as he cannot possibly refrain from keeping his mind at work on his favourite subjects, this condition is difficult to enforce. His mind appears to be quite clear, though his articulation is a little difficult; and his interest in everything connected with geology is as keen as ever. He is actually, in his infirm state of health, working at a new edition of his "Antiquity of Man." Mary, notwithstanding her constant anxiety and cares for him, looks very well and is charming as ever.

I had much pleasant botanical talk with Katharine, who keeps up all her old interest in that subject.

———

October 1st.

My Barton rent audit—very satisfactory—luncheon to my tenants and Mr. Smith. Mr. Hutchings went away.

———

October 7th.

We returned on the 5th (Saturday) from a visit of two days and a half to the Millses at Stutton. It

1872. was a very pleasant visit, though the weather was abominable, and I was unlucky enough to be sharply attacked by rheumatism, which marred my enjoyment. But it was a great pleasure to see our dear old friends again and to find them fairly well in present health, though both very fragile ;—cheerful also, and as kind and cordial and pleasant as ever. Lord John Hervey was staying there, and was exceedingly pleasant ; and his cousin John, who was in great force, and whom I am always glad to meet, dined there on the Friday, as well as Mr. Zincke, Mr. William Gurdon, and Mr. Wood the curate.

On the 3rd, Mrs. Mills took us over to see the Bernerses at that beautiful place Woolverstone ; a furious storm which came on prevented us from seeing anything of the out-door gardens ; but we saw the conservatory, which is one of the most beautiful I know anywhere, both for its general effect and for the choice and the fine growth of the plants, especially of the climbers. An extension of it has lately been made, arranged so as to look like a grotto or rocky chasm, and planted with Ferns.

Sir Henry and Lady Thomson,* Mrs. Charles Berners, and Miss Maude Berners, staying at Woolverstone.

Saturday, 5th.

Very cold weather.

Still suffering from rheumatism, but we posted from Stutton to Hintlesham— had luncheon with the Anstruthers—then to Stowmarket, and so home.

* The celebrated surgeon and artist. His wife was a fine musician.

Monday, 7th October. 1872.

A beautiful autumn day.

A visit from Lady Cullum and Mrs. Flood.

————

October 10th, Thursday.

Saw a Vanessa Antiopa (Camberwell Beauty) in the pleasure ground—the first I ever saw here. The species is known to have been remarkably abundant in England this year.

══════

LETTER.

Barton, Bury St. Edmund's,
October 10th, 1872.

My Dear Mary,

What do you think I saw this morning in the pleasure ground?—a *Camberwell Beauty*—Vanessa Antiopa, a beauty indeed: the first I ever saw in England, though I have repeatedly seen it in Italy. You know that an unusual number of them have appeared in England this summer; *Nature* says, as many as 200 have been caught.

It is rather too late to wish you joy of your birthday, as I missed doing so on the correct day: but you will not doubt that I most heartily wish you happiness and many many happy returns of the day.

You are now, I suppose (or very soon will be), about taking leave of dear Susan; pray give her my best love and hearty wishes for her safe and pleasant journey.

H H 2

1872.   Our visit to Stutton last week was very pleasant,
though a little marred, as far as I was concerned, by
a sharp attack of rheumatism. I am beginning to
feel decidedly that I am growing old, and that,
at least in such a season as this, the fireside is my
proper place. I have, however, now thrown off my
rheumatism for the present.

I have been reading a good paper by Dr. Brandis,
on the Forests of India, published in " Ocean
Highways,"—Clement Markham's Periodical.

In Grote, I have come to the battle of Ægos-
potami and the complete overthrow of Athens, and
I really felt pain in reading of it (though I knew the
main facts perfectly well before), just as one does in
reading a novel or poem, when some interesting
hero or heroine meets with a great misfortune.

With much love to Lyell,

Believe me ever,

Your affectionate brother,

CHARLES J. F. BUNBURY.

JOURNAL.

Saturday, 12th.

Very bad weather.

Lady Cullum, Mrs. Flood and two Miss Floods
came to luncheon and spent the afternoon with us.
Mr. and Mrs. Percy Smith dined with us.

Friday, 18th October.

Talked with Elmer, and gave him a present:

visited the Boys' School. We returned home. 1872.
The afternoon and evening excessively wet.

October 19th, Saturday.

We returned yesterday evening from Mildenhall,
whither we went on Tuesday the 15th. The
weather was very unfavourable, but we went on the
16th with Scott and Betts to see the new plantations, which have been made in the last 5 years,
an additional piece of land being progressively
planted each year. They are of Scotch fir and
larch, with a smaller proportion of oaks, they look
very healthy and are growing very well. Great care
is taken to trench the ground thoroughly, and
especially to break up the sort of crust (which the
labourers call "iron-hard,") of sand partially con-
solidated by iron which is always found at a depth
of from 6 inches to 2 feet below the surface.

On the 17th we went with Betts to Holywell and
West Rows: saw some cottages, and, in particular,
inspected two farm-houses which are building (not
yet quite finished), for two small farms on the edge
of the Fen. These (which are something intermediate between cottages and farm-houses), appear
to me to be very good. All the timber used in
them was grown on the estate.

Weather excessively wet all this week. I
selected some parcels of dried plants to be sent to
Barton.

Sunday, 20th October.

Yesterday afternoon's post brought us the most
sad and distressing news of the death of our dear
friend Lady Campbell : she died in child-birth on
the 17th.   It is a most painful shock, and very
unexpected, for we had heard nothing of her being
in any danger.   We have yet heard no particulars.
She was a truly charming and most interesting
woman, admirably good and very attractive : we
both loved her very much.   To her poor husband
the loss is terrible, it must be an overwhelming
blow,—I dread to think of his desolation.   They
were devoted to each other—wrapped up in each
other : they had been married 19 years, and now he
is left alone, with the charge of 12 children, most of
them under 12 years old.   He is an excellent and
most deeply religious man, but it will be hard for
him to bear up under such a blow.

This is the second dear and charming friend we
have lost this year, through the dangers of
parturition—Mrs. Walrond being the first.

Monday, 21st October.

Wrote to Edward.   We arranged some books.
Octavia and Wilhelmina Legge arrived.

Tuesday, 22nd.

The Quarter Sessions at Bury.   Saw Clement in
his wig and gown.   Minnie Powys and John Hervey
arrived.   Mrs. and Miss Wilson dined with us.

Leopold and Lady Mary Powys, the Lynedoch Gardiners, Sir Edward Greathead, Mr. W. Gurdon, Admiral Spencer, Mr. William Hoare arrived.

———

Thursday 24th.

Read Horace, b. i., Odes 10-13. Read some more of Grote, and went on with Parlatore on Conifers.

All the party, except Fanny and myself, went to the Bury Ball.

———

Friday 25th.

Our dance very pleasant, and altogether a great success.—I danced Sir Roger de Coverley with Lady Cullum.

———

Saturday, 26th October.

The Powyses, John Hervey, Sir E. Greathead, Mr. Gurdon, Mr. Blakesley and Mr. W. Hoare went away. Dear Minnie arrived, and in the evening dear Kate Hoare and her husband. Lady Mary Hervey dined with us.

———

October 27th.

Fanny had a letter, a few days ago, from Lady Frere, confirming what we had read in the newspapers, that Sir Bartle Frere has consented to go out to Zanzibar, as an envoy from our government to the Sultan of that country, to negotiate a treaty

1872. for the suppression of the East African slave-trade. It is very noble, in a man of Sir Bartle's age and station and fame, after a life of labour, and after holding such high and honourable offices, to give up ease and comfort and domestic enjoyments, and undertake a mission to a barbarian and unhealthy country, solely for the benefit of an oppressed race. I admire and venerate him profoundly.

On this day, Sunday, we went to morning Church —a large party.

A very fine day—a rare advantage in this excessively wet and stormy season. Since the beginning of this month there have been very few days without heavy rain. Frost, however, there has been hardly any, not enough to cut off the Castor-oil plants (Ricinus), Aralia papyrifera, lemon-scented Verbena, and other half-hardy plants. The autumnal colouring of the trees is now in great beauty: the golden-yellow of the Sycamores, Horse-chesnuts and Elms and the orange-brown of the Beeches, very rich. Acer rubrum, too, is beginning to show beautiful colours, but the Liquidambar does not appear to advantage this season.

LETTER.

Barton, Bury St. Edmund's.
October 28th, 1872.

My Dear Henry,

It has made me very happy indeed to hear Sir James Paget's favourable opinion of your case, and I trust that we may now look forward with hope to many years of (at least) comparative health and 1872. comfort for you.

I do indeed feel very thankful for this. I have not written to you since the sad news of dear Lady Campbell's death ; a most melancholy and lamentable loss indeed, and a grievous shock to one's feelings. She was so charming a person and so admirable in every way, that she must be long missed and regretted by every one who knew her ; but above all we must feel for poor Edward, who was so devoted to her, and to whom the loss is quite irreparable—a crushing blow.

We had made all the arrangements for our parties before we heard of her death, so that we could not without great inconvenience, have put them off, which otherwise Fanny was much inclined to do. We have now a very pleasant party in the house, in particular, the two charming brides, Kate Hoare and Sarah Seymour, both looking very pictures of happiness.

Here are also your friends the Lynedoch Gardiners, both of whom I like very much ; and they fully sympathise in our joy at Sir James Paget's report of you.

I heartily hope Willy will go on to your satisfaction at Worthing.

We have now had two beautiful days after the wettest and most stormy October, almost that I can remember here.

Ever your very affectionate brother,

CHARLES J. F. BUNBURY.

JOURNAL.

1872. Most of the party were photographed.

We planted a tree for Sarah and Lord Albert Seymour:—they went away to Riddlesworth. Miss Isabel Hervey and her brother arrived.

_____

The Barnardistones arrived. Mr. and Mrs. James, Freda Broke, Captain and Mrs. Ives, dined with us.

_____

A very fine day.

Had a pleasant walk with Minnie and Lady Florence. All the party except Kate Hoare, Minnie, Fanny, Edward and I, went to the Thetford Ball.

_____

Dear Kate Hoare and her husband went away.

Charades very prettily acted in the evening by the young ladies, and John Hervey and his cousin William.

===

LETTER.

Barton, Bury St. Edmund's.
November 1st, '72.

My Dear Lyell,

The _Camberwell Beauty_ was, when I first saw it, sitting on the ground, on a corner of the lawn, at the foot of a mass of ivy, and I thought it 1872. might have been sucking the ivy flowers, as the Admiral is so fond of doing. After I had looked at it for a few moments, it flew swiftly away, and I lost sight of it. The border of the wings was certainly cream-coloured : this I observed because I had seen this peculiarity of the English variety remarked in books, but I cannot remember having observed the yellower colour in those which I saw formerly in Italy. The sudden appearance of comparatively great numbers of certain insects, at uncertain intervals, and without any obvious determining cause, is very curious.

Did you happen to observe in *Natnre* of October 10th, a short but important paper by Carruthers, on the Tree Ferns of the Coal Measures? What struck me particularly is, that he says he has been able to ascertain, that the fruit of Pecopteris arborescens (one of the commonest Ferns of the Coal formation), is so like that of recent Cyatheas, that they cannot be *generically* separated. I am not at all surprised at this, as the similarity of the fronds is most striking, but I have never seen fructification of that Pecopteris shewing structure. It is a very interesting and striking instance of the persistence of some generic types. Truly the pedigree of Cyathea is of venerable antiquity. You know, no doubt, the magnificent stems of Cyathea in the British Museum. And Carruthers seems to think that the stems of those Carboniferous Cyatheas were *Psarolites*.

Ever yours affectionately,
CHARLES J. F. BUNBURY.

JOURNAL.

1872   Our charming party broke up, all went away except Minnie and Edward.

Dear Sarah Seymour arrived.

I received a very agreeable letter from Sarah Hervey.

Received the "English Botany" (the original edition), which I had bought from H. Sotheran for £22 10. o.; a copy in good condition, the 36 volumes bound up in 23. It is a fine old classic on the subject, and in turning it over I recall the keen delight I used to have in it when it was lent to me in my boyhood.

We went to morning Church with Minnie, Sarah, and Edward and I received the Sacrament.

We have had our usual October gathering of friends young and old; and as usual it has been exceedingly pleasant; our dear Millses indeed were absent, as Mr. Mills's health is too feeble to allow of his leaving home; but otherwise we may well be thankful to have seen so few gaps in our circle.

There were the three charming girls whom I have so often mentioned,—Minnie Powys, and Octavia and Wilhelmina Legge; and dear Kate Hoare, whom it was an especial delight to see here with her 1872. husband, both looking very pictures of happiness; she is in brilliant good looks, and as delightful as possible. But I regretted the absence of Sarah Hervey, though I knew that she was well and happy. Our darling Sarah Seymour was not able to come in time for the ball, and only appeared among us with her husband for one day; but she has now come back alone to stay quietly with us and her mother for this week.

There were also Miss Isabel Hervey and her brother William; John Hervey and Lord John; Admiral Spencer; Colonel and Mrs. Lynedoch Gardiner (both of them great favourites of mine); and Sir Edward Greathead, a new acquaintance whom we found very agreeable. Altogether, it was a party of a very high order of agreeableness.

Edward went away.

Tuesday, 5th November.

Had a pleasant walk with Sarah; Minnie and Fanny having gone to call at Ashfield,—the evening very fine.

LETTER.

Barton, Bury St. Edmund's.
November 9th, 1872.

My Dear Katharine,

You will have heard from Fanny of our fortnight of gaieties connected with the Bury ball,

1872. and our own dance.   It was as pleasant as possible,
—a most agreeable set of guests, without a dis-
cordant or uncongenial element.   And this week
has been one of not less enjoyment, though quite
contrasted, for we have been perfectly quiet, without
any company except dear Minnie and her daughter
Sarah, both of them delightful.   Sarah is looking
remarkably well, and is everything that is charming,
quite as charming in this quiet life as in the most
lively soicety ; just as simple and loving and delightful to us as ever she was.   Her husband, poor man,
has not been able to come to us, being on duty at
the dépôt at Warley, where he sees nobody but raw
recruits.   However, we hope to see him early next
week.

I have made an addition to my botanical library :
English Botany, the original edition, the 36 volumes
bound up in 23, bought for £22 10s.   It is quite
a classic in our science, and in looking over it I read
vividly the pleasure I used to have when a boy, in
studying it, when it was lent to me a volume or
two at a time.   I have also ordered, but not yet
received, one of Jacquin's splendid works on exotic
plants from Quaritch.

I have a fact to tell you about a Moss.   I have
found Hypnum abietinum rather plentifully near the
Gravel pits here : the curiosity of the thing is, that
it grows on ground which I knew every foot of,
and which I explored botanically scores and scores
of times in my young days, when I was as keen
after Mosses as I could be, and much sharper
sighted than I am now, and yet I never found a bit

of this Hypnum at Barton. It must have es- 1872.
tablished itself since then, but *how* is the puzzle, for
it never fructifies in Britain.

We are both well and very happy—at least, I
can answer for myself.

The high winds have nearly stripped our trees
already, but there has not yet been frost enough to
cut down even the Castor-oil plants in the open
ground.

I hope you have had recent and good accounts
from Frank, and that all your children are well.
With much love to them,

Believe me ever,

Your affectionate brother,

CHARLES J. F. BUNBURY.

JOURNAL.

November 10th, Sunday.

We went to morning Church with Minnie and
Sarah : an excellent sermon from Mr. Percy Smith.

November 11th.

The Miss Richardsons arrived, and Edward.

The weather is excessively cold—storms of rain,
sleet, and snow.

November 12th, Tuesday.

Very cold and stormy.

1872.   Spent the morning in looking over sea-weeds with Joanna Richardson.

Lord Albert Seymour, John Herbert, Lady Rayleigh, and the John Paleys arrived.

Lady Hoste and Mr. Greene, the Wilsons and many others dined with us.

November 13th.

Lady Cullum and Mr. Beckford Bevan dined with us.   Lady C. very comical and entertaining.

Thursday, 14th.

Dreadful weather—rain, sleet, and snow.

Lady Hoste, with Mrs. Seymour, Miss Greene and two children came to luncheon and spent the afternoon here.   Miss Praed and Miss Thornhill arrived.   The Tom Thornhills dined with us.

Friday, 15th November.

Dreadful weather.

Many of our party went to luncheon at the Wilsons.   I looked over some more sea-weeds with Joanna Richardson.

Saturday, 16th.

Excessive rain.

All our party went away except Minnie, Sarah, Albert Seymour and Edward.

Again a week of company. The two Miss Richardsons, John Herbert, Lady Rayleigh, Mr. and Mrs. John Paley and Miss Praed: besides Minnie, the Albert Seymours and Edward. Edward dined at Hardwick.

November 19th.

Edward went away.

LETTER.

Barton, Bury St. Edmund's,
November 20th, '72.

My Dear Henry,

Many thanks for your letter. I am very sorry that you are not feeling well, but it is not surprising, for the weather for a long time past has been so bad, so wet and stormy, that it must have a depressing effect on every one who is not in very good condition.

We are now almost alone—no one with us but Minnie, and she, I believe, must leave us to-morrow, so that for a few days we shall be quite alone :—a rare event with us. We had exceedingly pleasant company at the time of the Ball and for some time after, and another set last week. It was a great delight to have Sarah with us, first for a week alone with her mother (Lord Albert being on duty at Warley),—afterwards with her husband and some other company. I was very glad to have the

1872. opportunity of improving my acquaintance with her husband, and I am very much pleased with him : he is not only good and amiable and of most pleasing manners, but very intelligent also, with much observation and desire of learning. I do really believe he is worthy of Sarah, which is saying a great deal. They both look as happy as possible. Sarah is quite as charming as ever, as sweet and bright and happy-tempered and kind and loving to us—as fascinating in every way as ever, and in beauty, I think she is even rather improved. It is delightful to see her so happy, and dear Minnie so quietly happy too.

I hope Willie will get on well at Worthing, and that you will find a satisfactory teacher for George.

With much love to Cissy and Emmy,

Believe me ever,

Your very affectionate brother,

CHARLES J. F. BUNBURY.

JOURNAL.

Saturday, 23rd.

We drove to Ashfield, and saw poor Lady Susan Milbank and her children.

November 25th.

Violent wind and storm.

Tuesday, 26th.

Mrs. Ellice and her two daughters, Mr. Goodlake, Mr. Zincke, Mr. Montgomery and Captain George Blake arrived.

Captain and Miss Horton, Evelyn Bevan and her brother, and the the Victor Paleys dined with us.

Wednesday, 27th.

A fine mild day. Had a pleasant walk with Cissy and Helen Ellice. The Abrahams, Lady Cullum and the Percy Smiths dined with us—pleasant.

Friday, 29th.

A fine day.

All our gentlemen went away. Mr. and Mrs. and Miss Fuller Maitland, Mr. Clarke, Mr. Pryor and Mr. Tyrell arrived.

Lady Cullum and Mr. and Mrs. Hardcastle dined with us.

Saturday, 30th November.

Had a walk in the rain with Cissy and Helen Ellice. The Maitland Wilsons dined with us.

December 2nd, Monday.

Again a week of company:—Mrs. Alexander Ellice and her daughters Cissy and Helen (these are with us still), Mr. Goodlake, Mr. Zincke, Mr. Montgomery and Captain George Blake (these four

1872. from the 26th till the 29th), Douglas Galton and his daughter Evelyn, Mr. and Mrs. and Miss Fuller Maitland, Mr. Clarke (of Trinity College Cambridge), Mr. Pryor and Mr. Tyrell, (all these came on the 29th and went away this morning).

Mrs. Ellice and her daughters (especially Cissy), are very clever and agreeable, good-natured and friendly. Mr. Goodlake has travelled a great deal, and seen many distant and interesting countries: indeed, he seems to have been in most parts of the world, he has therefore of course seen a great deal, he has observed much, and talks well of what he has seen.

Mr. Zincke (whom we have long known), is likewise a man who has travelled and observed much, a man of much cultivation and very active mind, lively, conversible and pleasant.

I see in the newspapers the death of Mrs. Somerville—"at Naples, suddenly, in the 92nd year of her age." So her noble life,—spent in the noblest and most elevating pursuits, has been closed by a quick and (we may hope) a painless death. She was not only a woman of extraordinary genius and attainments, but a most amiable and excellent one; simple, modest and gentle and unpretending to a remarkable degree.

I have been acquainted with her ever since I was a boy, and was much in her company when she was living with her husband and daughters at Chelsea, and I was in London in 1835, '36, and '37 ; and again when they were living at Rome in the spring of '43, I received much kindness from her, and enjoyed much of her easy, pleasant, cheerful and 1872. instructive conversation. Fanny and I saw her again at La Spezia in May, '66. She seems to have preserved all her faculties (except that of hearing), unimpaired to the close of her life, and last spring she took (as I heard) an eager interest in the eruption of Vesuvius. Mrs. Somerville's death (as we have since heard from Mary Lyell), was as easy as her best friends could wish. She died in sleep without a struggle ; and she had been out of doors the day before. It was such a death as was fit to crown her blameless and noble life.

There was a good notice of her in a letter to the *Times* by Sir Henry Holland.

Tuesday, 3rd December.

Mrs. Ellice and her daughters went away.

Thursday, 5th.

Very cold—a sharp frost in morning.

We went by railway to Mr. Fuller Maitland's, Stanstead House ; left Bury at 11.50, reached Stanstead station at 2.30.

Friday, 6th.

At Stanstead, Fanny went with Arthur to see Mr. Marshall's school near Hertford, and returned about 5. I stayed at Stanstead, walked with Mr. Maitland to see his new house ; he showed me also his pictures and books—very valuable.

1872.   Mrs. and two Miss Byngs and some other guests came to dinner.

———

We spent the morning in looking over Mr. Maitland's pictures and library.

———

We returned yesterday from a visit to Mr. Fuller Maitland at Stanstead, near Bishop's Stortford. Mr. Maitland is a man of much cultivation, as well as of cheerful and pleasant manners: and he has formed rich collections, both of art and of literary curiosities. His pictures are very numerous : many of them rare and curious specimens of the very early masters ; and what gave me more pleasure, a great number of delightful landscapes by *Crome, Callcott, Constable, Collins, Turner* and other English masters, as well as by the best of the Dutch.   Also a superb landscape by *Rubens.*   His library includes many rare and curious books, especially *first editions* of "Paradise Lost," of Milton's minor poems, of the "Faery Queene," of Drayton's "Polyolbion" (with very curious illustrations), of Herrick, and others which I cannot now recollect. Stanstead House is a good house, and well situated, and the country about it is agreeably varied in surface, as well as abundantly wooded, much superior to those parts of Essex more to the north and to the east.

But Mr. Maitland is building (and has nearly

completed) a much larger and finer house, in a still 1872.
better situation, on the opposite side of the valley.
with very agreeable views over fine slopes of park
and woodland. The elms are especially fine ; some
of them surpassing in size and grandeur of form
almost any that I have seen. The elm seems in
fact to be *the* especial tree of that neighbourhood.

Monday, 9th December.

Signed leases to Mr. Latham of two houses in
Pall Mall.

======

LETTER.

Barton, Bury St. Edmund's,
December 9th, '72.

My Dear Katharine,

I hear that you mean to set off for Berlin
before the end of this week, so I write to wish you a
prosperous journey and much enjoyment in the
military capital. I am sorry for your undertaking
such a journey in such a season, but I hope you will
not suffer from it, and at any rate you will not set
off while the weather is so bad as now. We had a
tremendous gale all last night, and it is still very
rough and stormy. I am glad that Leonard is
going with you.

I was very much interested by what Mary wrote
to Fanny about Mrs. Somerville's death. It is a
great comfort to know that her noble life was

1872. terminated by an easy and painless death. There
could hardly, I think, be a more admirable or a
happier life, or a more desirable death:—such a
death is really "kind Nature's signal for retreat."
I had no expectation of ever seeing her again in
this world, therefore I cannot feel the personal
regret for her that those may who have been in the
habit of seeing her. As one advances in life, one
must expect to see the old friends of a former
generation drop into the grave; but what is deeply
and bitterly painful is, when the young are snatched
away in all their brightness and hope. And we
have lost *so* many young friends this year! I feel
deeply thankful, however, for Kate Hoare's recovery.
I am very sorry for Martha and Mary Somerville,
whose home is so suddenly broken up, and who are
past the age when it is easy to begin a new career.

I have just got the last part of "Middlemarch,"
and have not yet had time to read it, but I have
peeped at the end, and see that it concludes
wrongly. Have you read that book? if not, I think
it would be a pleasant companion on your journey:
it is extremely clever and well written, and well
worth reading, though there is nothing in it so good
as Mrs. Poyzer in "Adam Bede."

You may have heard that we returned on
Saturday from a visit to Mr. Maitland at Stanstead,
and that they had very kindly asked us thither, in
order that Fanny might more comfortably make out
her visit to the school near Hertford, to which
Arthur MacMurdo is to go next month. Her visit
to the school was very satisfactory, and I spent

the day pleasantly with the Maitlands, who were 1872.
very kind, and have a very remarkable collection of
pictures, especially a multitude of beautiful landscapes, both of the English and Dutch Schools,—
also many rare and curious books.

Now we are alone, and I hope we shall be quiet
till after Christmas, for there are many things I
want to read.

Believe me ever,
Your affectionate brother,
CHARLES J. F. BUNBURY.

JOURNAL.

December 12th.

Received Jacquin's "Icones Plantarum Rariorum"
3 volumes, folio, a fine work, bought for ten guineas
from Bernard Quaritch, Piccadilly.

Saturday, 14th.

Received a charming letter from Sarah Hervey,
and Fanny received one from Mrs. Kingsley. A
visit from Mrs. Hardcastle. Began to make notes
from Jacquin, and went on with annotation of
English Botany.

Monday, 16th December.

Dismal weather.

Sent money to Mr. Gedge for Mildenhall
Charities, and wrote to Mr. Prigg about some

1872. antiquities.   Began to make a list of our trees. Examined and arranged some minerals.

Visit from Mr. Percy Smith, who stayed all the afternoon.

––––––

Wednesday, 18th December.

Sent   money   to   Miss   Bucke   for   Mildenhall Charities.

══════

LETTER.

Barton, Bury St. Edmund's,
December 18th, 1872.

My Dear Henry,

The rain it raineth every day, and nearly all day long, and there is no sign or hope of any change for the better.   If this goes on much longer, we must become amphibious and web-footed : indeed such a change appears inevitable, on Darwinian principles, to adapt us to the changed condition of things around us.   I never remember such a season ; the winter of '52, in which you were married, was something like it, but hardly so bad, I think.   I am afraid this weather must be very bad for your health, and I am not surprised, though very sorry, to hear that you are suffering.   One ought to be very strong, not to feel the depressing effect of it. I am pretty well, but rather *flabby*, both in mind and body —wet-brown-paperish.   Fanny complains of headaches from the same cause.   Arthur alone feels no depressing influence, but is full of glee and

joyousness all day long, preparing for Christmas. 1872.
Most heartily do I wish you and dear Cissy and all
your children a happy Christmas and New Year,
and all the happiness that a new year can bring. I
hope you will soon be successful in finding a good
tutor for George : it is a most important question,
and I well understand your anxiety, as the choice
of a tutor may have such an important influence
upon his intellectual progress and his future career,
at the same time considering health : I am afraid
there may be a good deal of difficulty in finding a
suitable teacher for him. I am glad to tell you that
none of our more valuable trees were at all damaged
(to speak of), in the gale of last Sunday week,
though it roared tremendously : only an old red
Cedar, in a bed opposite to my window here, was
torn asunder, and one half blown down (it divided
into two main stems from very near the ground), it
is no great loss, however.

We dined yesterday at Stowlangtoft—a good
party; I sat between Mrs. Wilson and Mrs. Browne,
the Bishop's wife, both very pleasant.

With much love to dear Cissy, Emmy and the
boys,

Believe me ever,
Your affectionate brother,
CHARLES J. F. BUNBURY.

JOURNAL.

1872.   A rather better day.
Went with Fanny to the School to hear the girls
sing Christmas hymns.

LETTER.

My Dear Mary,

Most heartily do I wish you and dear
Charles a happy Christmas and New Year—that
the coming year may be one of health and
prosperity and happiness to you, and may be
followed by many more as good. I hope you are
not feeling any ill effects from this most extraordinary season of rain and gloom : do you
remember anything equal to it ? I do not think
'52 was quite as bad, though in the same style.
It seems as if all the rain, in which several seasons
in the last ten years had been deficient, were now
paid to us at once. The worst part of the matter
is, that we are in serious fear of the fens being
flooded, which would be a heavy calamity : the
water in the river (the Lark), is up very nearly to
the top of the bank, which is many feet above the
level of the surrounding country, and if the bank
were once to give way to the pressure of the water,
thousands of acres would be drowned. It is

Lombardy on a small scale. The steam-engine in 1872. Mildenhall Fen is kept at work night and day, with great expenditure of coal, yet it can hardly keep up with the increase of the waters. This wet season does not seem to be bad for the health of the population generally in the country. I hear that Bury, and the country hereabouts in general, are unusually healthy. But it makes us individually, me at least, feel very weak and *flabby*—wet-brown, paperish. Arthur, indeed, is anything but depressed, he is in immense spirits, and full of eagerness in arranging Christmas decorations and preparing for Christmas festivities.

Fanny, I need not say, is very busy. Since we returned from Mr. Maitland's, I have had a nice quiet time for reading. I am still engaged with Grote, and am now in his tenth volume, and just coming to Epaminondas : I feel very bitter against the Spartans. I am reading Geikie's little book on the " Scenery and Geology of Scotland," and am very much pleased with it :—also reading Lecky's " Leaders of Public Opinion in Ireland," which is very interesting and very agreeably written. I have lately read over a part (the section relating to Italy), of Hallam's " Middle Ages." What an admirable writer he was !"

Did Lyell happen to observe in *Nature*, of October 10, a short notice by Carruthers on Fossil Tree Ferns, which struck me a good deal ? What I thought striking was this :—he says he has been able to satisfy himself that *Pecopteris arborescens* (one of the commonest fossil Ferns of the Coal

1872. formation), is not distinguishable *generically* from the recent Cyatheas. I can readily believe it, though I have never myself happened to see a specimen of that Pecopteris in which the structure of the fruit could be made out : but it is curious and interesting to see the same forms of vegetation continuing through such an enormous lapse of time.

With much love from both of us to yourself and your husband,

Believe me ever,

Your very affectionate brother,

CHARLES J. F. BUNBURY.

JOURNAL.

December 24th, Tuesday.

Received a most charming letter from Sarah Hervey. She is the most agreeable of correspondents.

Mr. Beckford Bevan came to examine a MS. in our library.

Walked about the grounds, the day being really fine.

Christmas Day.

A fair day.

We went to morning Church.

Thursday, 26th.

A beautiful day.

Went out with Fanny in the pony carriage—we were out nearly two hours.

Received an agreeable letter from Mary.

Arranged some Mosses.

Cecil, Clement, Herbert, and Charley Napier arrived.

The labourers' supper party.

———

Saturday, 28th.

A beautiful day, like spring.

Found a *buttercup* in flower.

The children's merry-making—the young Smiths and Scotts with Arthur—the Christmas *pie*.

———

December 30th.

Again we are near the close of a year, and are naturally impelled to look back on the course of it, and to review what we have done, felt, lost, gained, and learned in the time—"in that regretted time," as Charles Lamb says. And again I deeply feel, how much cause I have for the most earnest and heart-felt gratitude to God for all His goodness to me. I can only repeat what I have written again and again in previous years, as to the many causes I have for deep thankfulness. I feel all this as deeply as ever, notwithstanding that a sadder shade has been cast over this year by anxiety about my brother, and by the loss of some dear friends. In the early part of the year, Henry's state of health was such as to cause the deepest anxiety, and indeed the most terrible apprehensions. Sir James Paget's skill relieved him from the immediate

1872. evil, and us, for a time, from apprehension, but I
much fear that now alarming symptoms are
gathering.—God grant it may be otherwise.

As one advances in life (I am now well on in my
64th year), it is natural and inevitable that our
friends of the older generation—those who were in
mature life when we were children, should drop into
the grave : and though one may regret them, one
feels that there is no reason to repine or complain.
Thus, the death of our dear friend, Lady Napier
(at the age, I believe, of 87), and that of Mrs.
Somerville (at 92), were in the course of nature,
to be expected, and gave us no shock.   But
it is a very different thing when the young are
snatched away in their bloom and brightness from
the circles they were adorning, from the friends who
had hoped to be gladdened by their presence for
many a year.   In this year we have lost three dear
young friends, (perhaps I should say four, though
Ella Kennaway was not so intimate as the others).

Theresa Boileau, indeed, had for some time been
in such a state of health, that we were prepared
for her death, and when we saw her in April, I felt
sure that she could not live through the year,
though I hardly expected her departure to be so
speedy as it proved.   But those two charming
beings, Lady Campbell and Mrs. Walrond, were
snatched away quite unexpectedly, and we can
never cease to remember them with loving and
longing regret.   On the other hand, this year has
been marked by two joyous events : the happy
marriages of Kate Hervey and Sarah Napier, both

very dear to us ; and both marriages seem likely to 1872.
fulfil all their best promises. Both girls seem to
have married men who really appreciate them, and
who are as *nearly* worthy of them as men can be
expected to be. My wife and I have both been
blessed with good health during this year, for which
we may be very thankful. We have also, in this
year, lost two pleasant neighbours, though I should
hardly go so far as to call them *friends:*—Sir
Edward Gage and Mr. Milbank. The latter I am
particularly sorry for, because I have a great regard
for his wife. I have forgotten to mention another
old friend who has likewise departed within this
period—Lady Smith (Sir Harry's widow) whom I
knew so well at the Cape, and with whom I have
kept up an agreeable acquaintance ever since she
settled in London.

We have not travelled much, but have gone a
little further-a-field than we did last year, and I
gained one new idea by seeing Dartmouth.

We have enjoyed a great deal of agreeable
society, both here and in London,—I mean, of such
society as leaves a permanently agreeable impression
on the memory: this I have recorded from time
to time in my journal. As regards studies, this
year has perhaps not been as well occupied as some
previous ones, but perhaps it is only that my
reading has been less varied, my attention having
been a good deal concentrated on Grote's " History
of Greece." That is a long work, and I am not
even yet very near the end of it, but it is worth the
time and worth reading carefully. I have written

1872. nothing, except this journal and some botanical notes and sundry letters, my favourite correspondents being Sarah Hervey and Katharine and Mary Lyell.

December 31st.

This last day of the year like nearly the whole of this month, remarkably mild.   Indeed, though there was very chill ungenial weather in September and the early part of October, there has been no real *winter* cold at all,—hitherto, the mild temperature of December has been extraordinary, thermometer very seldom as low as freezing point, and the little yellow Aconite (Eranthis hyemalis), quite in bloom before the end of the year.   But the excessive and almost continual rains throughout October, November and December, have been still more remarkable.

So ends 1872.   God grant that the coming year may be better spent.

FINIS.